2020s

& THE FUTURE BEYOND

2020s & The Future Beyond

2020s

& THE FUTURE BEYOND

How To Survive Technology's Disruption in the Coming Decades – Winning the War Against AI, Robots, and Machines.

Kelly Idehen

DISCLAIMER

The content in this book has been provided as a detailed guide for the reader on what to expect and how to prepare for the technological disruptions in their workplace and communal life in the coming decades (2020s and beyond). It provides the reader with a roadmap on the technologies and trends to watch out for and how both corporate businesses and the government need to participate in this discussion for a sustainable future for everyone.

This book provides a general view of some fundamental debates we will have to face in the future, so it gives the reader time to do research and form their opinion. It further gives a thorough view of technologies such as artificial intelligence, nanotechnology, quantum computing, and other technologies so that the user can have an informed view of the opportunities and threats each of these technologies present.

The information contains fact and citations which are accurate; it also contains views expressed by the author based on his width of experience and research. However, it is not an exhaustive treatment of the subject matter, and experts' opinions may differ. Forming opinions based on the ideas contained in this book means taking full responsibility for choices, actions, and results. The author does not assume and with this disclaims any liability to any party or any loss, damage, or disruptions caused by errors or omissions resulting from accident, negligence, or any other cause. Therefore, the author does not accept responsibility for any results.

Copy editing by Midas Touch Literary Services
Cover design by Raphael Ben and Company.

To every conscious, human intelligence: who has taken a step back from the continuously turning wheel of life in search of the what, how, why and where of existence. May we become part of that transcended throng whose conscious presence will forever stretch beyond the confines of eternity.

2020s & The Future Beyond

TABLE OF CONTENT

INTRODUCTION

UNCERTAINTIES FOR THE FUTURE

~ Going from the Knowledge Double Curve, it took about 25 years for knowledge to double in size in the 1960s. According to prediction from IBM, in the 2020s, we will live in a world where knowledge doubles in size every 11 to 12 hours.

~ According to the research firm Gartner, the early 2020s will see the Internet of Things (IoT) grow to an estimated size of over 20 billion actively-connected physical devices sending, receiving, and processing streams of information.

~ According to Gartner, by 2022, 70% of enterprises will be experimenting with immersive technologies for consumer and enterprise use, and 25% will have deployed to production.

~ In the coming decade, 2020s, Japan plans to become the first nation to build a robotic lunar outpost, this outpost in the moon will be built entirely by robots and solely for robots.

~ The number of Internet users worldwide is projected to reach 5 billion in the 2020s; this will mean more than half of the world population will be connected to the Internet for the first time.

~ In the coming 2020s, the world of medical science will make some significant breakthroughs. Through brain implants, we will have the capability to restore lost memories.

~According to a report by Mckinsey, "62%" of executives believe they will need to retrain or replace more than a quarter of their workforce between now and 2023 due to automation and digitization.

~ By mid-2020s, 3D printing will gain a significant milestone. It will become faster and cheaper (near zero-cost) to make items such as clothes. A 3D printing process that took an average of 4 hours in 2014 will take less than 8 minutes to process in the 2020s.

~ The 2020s will provide us with the computer power to make the first complete human brain simulation. Exponential growth in computation and data will make it possible to form accurate models of every part of the human brain and its 100 billion neurons.

~ The prototype of the human heart was 3D printed in 2019. By the mid-2020s, customized 3D- printing of major human body organs will become possible. In the coming decades, more and more of the 78 organs in the human body will become printable.

~ The 2020s will see the emergence of new kinds of jobs and job titles. Some emerging job titles may include the following: Memory Augmentation Surgeon, New Science' Ethicist, Vertical Farmer, Waste Data Handler, Old Age Wellness Manager / Consultant Specialist, Virtual Lawyer/Teacher, Body Part Maker, etc.

~ Statistics found on Statista.com suggest that in 2016, there were only about 500,000 3D printers in the world. However, as we enter the 2020s, the number is estimated to grow to over 5.5 million units worldwide.

~ By the 2020s, Russia and the European Space Agencies (ESA) plan on 3D printing bases on the moon using moon soil instead of sending all the materials from the earth. This is a first of its kind and will open up space exploration like never before.

~ According to forecast made by Futurist Thomas Frey – By 2030 (approximately 12 years from now), up to 2 Billion jobs may disappear, which is like 50% of all jobs in the world taken over by smart robots and machines.

~ By 2025, half of the people with a smartphone without a bank account will use a cryptocurrency account.

~ 30% of world news and video content will be authenticated by blockchain by 2023 to prevent deep fakes.

~ By 2024, AI will be able to gauge your emotion before serving you an online Ad.

~ By 2023, 40% of professional workers will orchestrate their business applications, similar to how they manage music streaming.

~ According to a recent Ericsson Mobility report, in 2024, 5G networks will carry 35 percent of mobile data traffic globally. The report also states that 5G can cover up to 65 percent of the world's population in 2024."

~ According to Consultancy Accenture [81% of executives it interviewed think that within two years, AI will be working next to humans in their organization as "a co-worker, collaborator and trusted adviser"].

~ According to Jeremy O'Brien, physicist and professorial research fellow at the University of Bristol: "In less than ten years quantum computers will begin to outperform everyday computers, leading to breakthroughs in artificial intelligence, the discovery of new pharmaceuticals and beyond. The high-speed computing power given by quantum computers has the potential to disrupt traditional businesses and challenge our cybersecurity."

~ According to a report by Mckinsey, Self Driving Vehicles presents revolutionary change, its adoption will be evolutionary. We expect Level 4 autonomy—operating within virtual geographic boundaries—to be disruptive and available between 2020 and 2022, with full adoption coming later. Full autonomy with Level 5 technology—operating anytime, anywhere—is projected to arrive by 2030 at the earliest, with greater adoption by that time."

~ Owing to its wide range of uses, the global nanotechnology market is expected to grow at a CAGR of around 17% during the forecasted period of 2018-2024. In 2018, the global Healthcare Nanotechnology market size was 160800 million US$ and it is expected to reach 306100 million US$ by the end of 2025.

~ According to Statista, the augmented and virtual reality (AR/VR) market amounted to a forecast of 18.8 billion U.S. dollars in 2020 and is expected to expand drastically in the coming years.

~ Estimates suggest that in 2019, sales of virtual reality (VR) headsets will reach around seven million units, while augmented reality (AR) headset sales will climb to about 600 thousand. Forecasts project massive growth in both AR and VR headset sales in the coming years, with both technologies expected to sell over 30 million units per year by 2023.

~ According to Market Watch, the global CRISPR and CAS Gene market is expected to expand at a CAGR of 20.8% during the forecast period (2018–2026), owing to increasing research & development and demand for research.

~ 66% of customers are interested in buying items via VR. 63% of consumers said they are expecting VR to change the way they shop.

~ According to Goldman Sachs, consumer demand for drones will continue to build. By 2020, we expect 7.8 million consumer drone shipments and $3.3 billion in revenue, versus only 450,000 shipments and $700 million in revenue in 2014. By 2021, the commercial drone industry will have sold 1,000,000 units. Looking at the growth between 2018 and 2024, unit sales will have tripled in this time period.

****

As I sit to write this book, I have to choose between being seen as either a fanatical pessimist or a bigoted doomsday prophet – none of which I believe is true. As a futurist, it is my burden to continuously look towards the future through the lens of the present while standing on the summits of the past. As every futurist will know, the future is an ever-changing picture reeled on the films of conjecture. Given all the trajectories we see today, what does the future hold? What becomes of us as humans? Where will the balance of our socioeconomic wheel shift to?

From economists to technologists to physicists to religionists, all have sought to chart a path for the ship of humanity and give a forecast on what shores it would eventually wash-up on. But if history has taught us one lesson time and time again, it is that the future has a mind of its own. Like a river, it snakes its way through time, choosing what rocks to erode and what beach to flood.

When man first invented electricity, little did they see to what scale their invention will change the course of history. Today, without electricity, life will be at a standstill. We have electricity running our cars, our homes, our industries and we have electricity as an integral part of our ability to communicate. Electricity opened up new opportunities for work; electricity is instrumental to the socioeconomic progress we have enjoyed over the last 200 years. Electricity belongs to that class of innovation that is regarded as a General Purpose Technology (GPT) – a technology that not only drives other technologies but also has the power to disrupt the economy and job market at a monumental scale.

Another modern GPT that has had as much impact on humans as much as electricity had is Information Technology. The advent of computing devices, the Internet, and communication processes have revolutionized the meaning of life and extended the possibilities of what humans can do in this century. Information and Communications Technology (ICT) has created new kinds of jobs and become an enabling layer for many new kinds of inventions and innovations. From politics to health to agriculture to education, we find that ICTs have infiltrated and copulated with every sector, giving birth to a new identity and innovation within each industry.

The question then is, is human civilization now due for a new kind of General Purpose Technology (GPT)? One which, like Aladdin's magic carpet, can take us on a ride into the future? I believe so. I believe we are at the threshold of entering into a new era, one where artificial intelligence (AI) will act as a GPT and drive human civilization through into a new dawn.

So why AI? Why should AI stand out in an era where numerous other technological grounds are being broken? To explain this, we need to start our GPT narrative from electricity. Electricity is essentially the movement of electrons within a conductor. With our ability to transfer signals (energy) from one point to another, we could then communicate through digital electronics while effecting changes in our physical world. Then came ICTs, which helped to broaden the scope and functions of what electrical energies can help us accomplish. We have now digitalized almost every aspect of our lives; we can control the movements of these electrons, which mean we can organize information better and give machines the ability to talk to other machines. In other words, both electricity and ICTs helped us to improve communications between ourselves and our physical world. You can think of electrons as the cells and electricity as the blood. Both the cells and the blood flow in a network of veins, tissues, and organs, which we will regard as the different infrastructures and networks that make up the Internet and ICTs in general.

If we build AI on top of these previous layers as we are beginning to do now, it will mean we are giving these electrons the power to organize themselves, to form thoughts and meaning for themselves, to communicate with us, and between themselves, to use the energy provided by electricity to effect change both in our physical world and in the digital world. AI becomes the unconscious brain (unconscious until the idea of the Singularity becomes a reality) that can control how this network of electrons and energy effects change both in the digital world (where most of us currently live) and in the physical world where we all live.

AI will not just stop here. It will move on to enhance other technologies such as virtual reality (VR) and augmented reality (AR), to blur the line between the old era (dependent solely on our physical world) and the new era of mixed reality (one where there is no clear separation between the digital and the physical). AI is like that technology that enables us to breathe life into our everyday appliances. Our cars, our homes, our kid's toy, our phones, and every single thing around us will spark up with a life of their own when touched by the breath and power of AI.

AI, as it advances into Artificial General Intelligence (AGI) and Artificial Super Intelligence (ASI) may not end up becoming the last GPT for human civilization, but it is regarded by many technologists and futurists as possibly the last significant invention of man. AI has the power to keep improving. It has the unique ability to keep iterating over itself, to keep searching for answers over the vast network of information it controls. For the first time, humans will give power and life to their invention and let it chart the next course of their journey. From self-driving cars to the many automated production processes, we will end up creating; AI will drive us into the next era of human civilization. We will allow the creation to create, and according to futurists and technologists' world over, there is only one likely path where this road will lead to – the Singularity (the point where computer intelligence surpasses human intelligence).

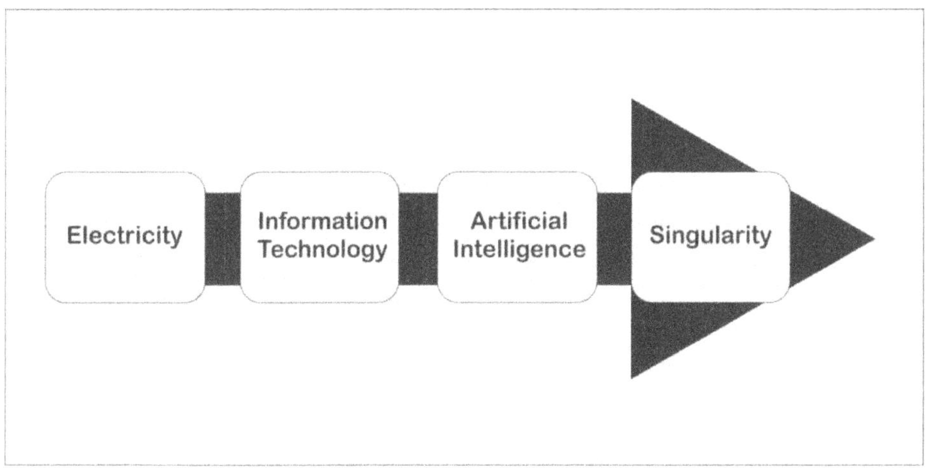

Progress of General Purpose Technologies (GPTs)

It is not my intention to describe futuristic ideals or detail a roadmap for what the future would be. We are at the turn of a new decade, a decade which I believe will usher humans their last opportunity to stand against a crisis that is even now at their gates. The journey of AI has only just begun, and we need to stand up to the resultant economic, political, and philosophical fallouts.

For example, I once gave a lecture to a gathering of young religious people; the presentation was a Christian perspective on the need to prepare for a world where technology takes center stage. In the lecture, I presented four main foundations that may redefine life in the future: Advancement in quantum computing, the pervasiveness of artificial intelligence, the spiritualism involved in Trans-humanism, and the possibility of a world where machines reach the Singularity. Well, after showing dramatic videos of Boston dynamic's robot doing backflip and Hanson's robot, Sophia, giving speeches and interacting almost like a sentient being, a young girl came forward with her question. She asked if God will destroy the robots when He comes.

As naïve as you may want to think of her and her question, it put me in a very uncomfortable position. I could not provide a satisfactory answer at the time. Looking back now, I realize I could not provide a clear-cut answer at the time because my answer will need to be nuanced to her experiences for it to make sense to her. Since then, I have had to think about the billions of Christians worldwide who share in almost the same fundamental faith. I have thought about the Moslems, the Buddhists, the Hindus, and every other faith-based religion; how would these religionists rationalize questions for answers that AI will give.

When nanotechnology, CRISPR, and biotech gives us the power to set new definitions for mortality, what answers will the religionist give? When through augmented reality and virtual reality, we set new definitions for what vision and illusion should mean, what will the religionist answer? What will be the religionists' interpretation, when Transhumanism built on-top of the narrative of evolution as opposed to Creationism, ushers humans a more evolved biology and body – Human 2.0? More importantly, how would religionists rationalize the reality of a conscious intelligence (the Singularity) more superior than them, would they think it a manifestation of their God, or find it an abomination to be disparaged at all cost?

These are some of the questions society would be forced to address objectively as opposed to subjectively within the coming years. However, even as religionists and those concerned with the bigger question of human consciousness grapple with these new questions of eschatological significance, politicians, world leaders, government, and corporate industries also have their existential demon to surmount when this clash of humans against machine intelligence becomes pervasive.

As the tech billionaire, Elon Musk once boldly put it, "***With artificial intelligence, we are summoning the demon.***"

The greatest of this demon is the two-headed monster of massive job displacement and the rise in inequality. Unlike the other General Purpose Technologies (GPTs), which allowed us to do work more efficiently and at scale, AI, especially when coupled with automated robots, is quickly becoming more intelligent and efficient at jobs that required human input. With modern progress in AI development, we are giving computers and robots the power of sight, sound, speech, and the intelligence to use these in becoming autonomous. They will flood our homes, they will flood our places of work, and they will relieve us of tedious and repetitive tasks. Nevertheless, as AI and automated robots count their wins, the humans whose job they have usurped would be counting losses.

As these humans lose their sense of worth alongside their means of livelihood, they would sink lower below the already unfair socioeconomic ladder. Unfairly enough, on the other hand, those who have the power to muster these robots to their aid (entrepreneurs and business owners mostly), will be given insane boosts up the socioeconomic ladder. This naturally results in imbalance and unfair class-divide between the rich and the poor, the haves and the haves not, those who are creating value in the robot economy and those who may become valueless and dependent on aid.

Like the steam engine and electricity, which drove the industrial revolution of the 80s and ICTs, which drove the information revolution of the 90s, AI will lead other technologies like quantum computing, VR, nanotechnology, etc. to drive a new revolution in the 2020s. However, unlike the previous revolutions of the 80s and 90s, this one will have the power to disrupt the very foundation of our society and weave a different kind of future, one which the best economists and futurists can only now be uncertain of.

I do not intend for this book to sensationalize any form of idealism for the future. It was birthed out of the questions we will be forced to answer within the next few years. I believe our duty just before we cross this definitive line into the 2020s, is to accept and begin preparation for a world where the unknowns and the uncertainties become the constants and the consistent(s). So journey with me through the chapters of this book as we make twists and turn through the philosophical and socioeconomic maze the 2020s will set before us.

PART ONE

PREPARING FOR THE FUTURE

1. TECHNOLOGIES THAT WILL SHAPE THE FUTURE

Cars that drive themselves are no longer sensational news today. The news only becomes interesting when it is a flying car with self-piloting/driving features, but even this grand feat, once attained, will soon cease to thrill us as we march decisively onward to a future where technology erases the line between our imagination and our reality. Today we are flooded with too many smart devices and technological breakthroughs that it is almost impossible to keep up with all of the trends. Will this upward technological climb continue? Should we be watching out for specific epochs in this transition of the human civilization? To answer this and many other questions, let us start our journey from Buckminster Fuller and his Knowledge Doubling Curve.

Buckminster Fuller was a renowned inventor and futurist of the 20th century. After extensive research and analysis of the progress made by human civilization, he noted that as time progresses, the rate at which knowledge increases becomes exponential. According to his research, he noticed that in the 1900s, knowledge doubled every century. That is, it took over 100 years for knowledge to get a two-time increase in value. However, by the 1960s, knowledge was doubling every 25yrs. Today, on average, knowledge is doubling every 13 months. Going forward, according to an estimate by IBM, within the coming years, the build-out of the 'Internet of Things' will lead to the doubling of knowledge every 12 hours, that is, knowledge will double twice in only one day. Cars, houses, smartwatches, smart clothes, smart fridges, and just about anything that has a chip in it, will contribute to this knowledge avalanche in the future.

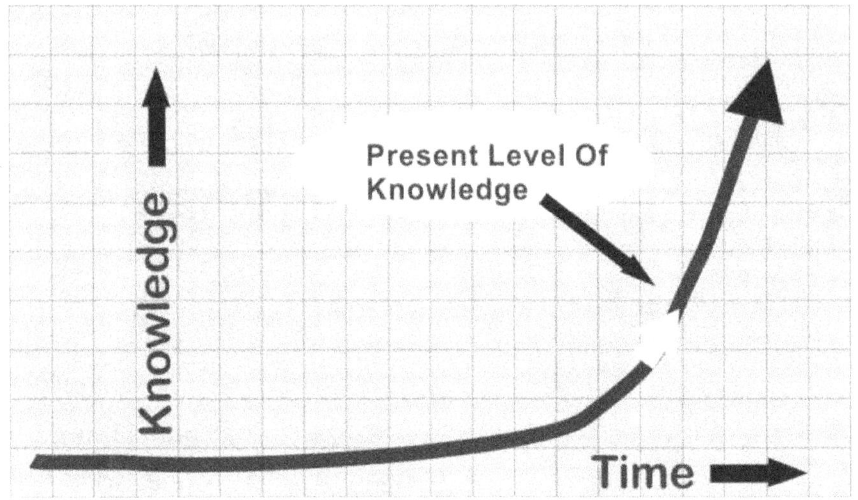

Exponential Acceleration of Knowledge as revealed by Buckminster Fuller's Knowledge Doubling Curve

In the future, it may be scary to fall asleep, because, before you wake up, the world may have advanced with new knowledge and trends leaving you behind (I am just being hypothetically sarcastic of course). However, more seriously, the question we must ask is, "Will technology keep growing, or will it reach a finish line in the nearest future?"

The knowledge doubling curve put forward by Buckminster Fuller is even more alarming when we juxtapose it with Moore's law, which has held for over 50 years now since the early days of semiconductor manufacturing. All of these provide the perfect stage for technologies, especially artificial intelligence, to drive civilization to what Ray Kurzweil propounds as the Singularity. Let us dig deeper into what these two renowned futurists, Gordon Moore and Ray Kurzweil, have to say.

GORDON MOORE:

Gordon Moore was one of the people who helped to make computers what they are today. He is one of the founders of Intel Corporation, a company that designs computer processors (an important component that functions as the computer's brain). Over time, as Moore kept working on developing the technologies that made processors and computer chips better, something extraordinary caught his attention. He noticed that every two years, the processor chip could take twice the amount of components like transistors - which meant that they could become two times better at their functionality. After years and years of observing this phenomenon, he published a paper on it, and the phenomenon became known as Moore's Law.

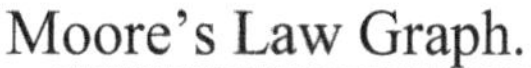
Moore's Law Graph.

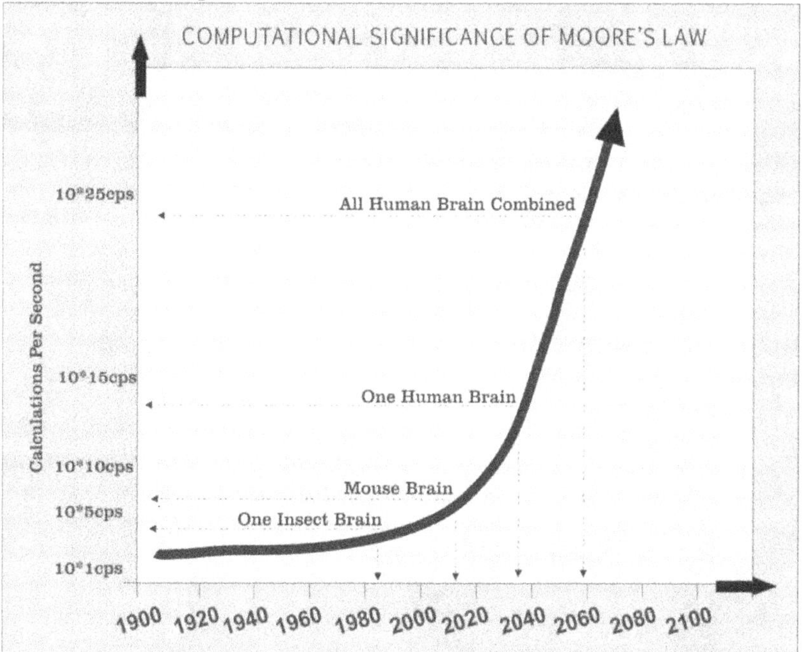

Moore's Law showing the accelerating rate of computation as more chips become smaller and cheaper

The law states that "***The number of transistors in a dense integrated circuit doubles approximately every two years.***"

The average smartphone today is millions of times faster than all the computing power of the computers NASA used for its Moon exploration mission in the 1960s.

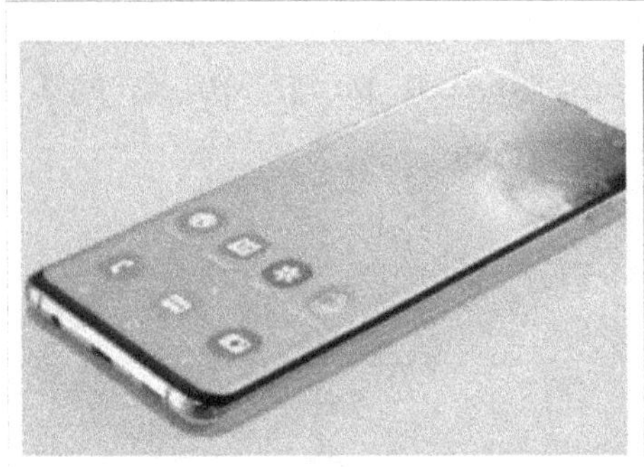

Today's Smartphone are millions of times more capable than the Apollo Guidance Computer (AGC) that helped NASA carry out their moon landing mission.

The effect of Moore's Law

In the 1950s, 5Megabyte of hard drive was such a valuable resource, that big companies paid thousands of dollars to rent it for months at a time. This 5MB of memory was so bulky; it was almost the size of a car, and it took over 12 men to install and uninstall as they moved it around.

It took over 5 men to move around a 5MB memory RAM in the 1950s, today a memory card that fits in your fingers can contain over 500GB of memory space.

Effect of Moore's Law

Today, the Samsung Galaxy s10Plus, a phone that is within your budget and can fit in your pocket, packs a RAM of 12Gigabyte, and an internal memory of 1Terabyte. All of these while having three different rear cameras and the ability to run on a 5G network. The phones of the future will offer even more capabilities than the Samsung Galaxy s10plus. Phone manufacturers, like Huawei, Apple & Google, through its Pixel phones, are continuously exploiting Moore's law and pushing the boundaries for the scale and power of computing.

As computer chips continue to obey the upward bend of Moore's law, we will continue to see mind blowing improvements in computing devices such as smart cars, smart clothes, smart homes, smart cities, and everything else that we decide to put computer chips into.

RAY KURZWEIL:

On the other hand, Ray Kurzweil, who won the National Medal Of Technology and Innovation Award in 2000, and currently works as Director of Engineering for Google, has a unique view of how technology will advance in the future. He believes technology will keep advancing until it gets to a point known as the Singularity. According to Ray Kurzweil, at this point, computers will surpass human intelligence (become smarter than humans), they (computers) will become self-aware, and these computers (whether in digital or physical robot forms) will keep improving themselves recursively.

In essence, computers are moving from basic artificial intelligence or Artificial Narrow Intelligence (ANI) to a level where we will say it has human-level intelligence or Artificial General Intelligence (AGI); and from there, computers will keep improving and eventually shoot pass humans to a higher level of Artificial Super Intelligence (ASI). A lot of people believe and forecast that when computers get to the ASI level, we will regard it as a form of god – for its powers and capabilities will be far beyond our human conceptions. I address this very controversial subject in more depth in the last section of this book.

According to Ray Kurzweil and other proponents of the Singularity phenomenon, there is a general belief that the Singularity – the point where computers attain human-level intelligence, or Artificial General

Intelligence (AGI) will occur somewhere around 2040 to 2050, which is roughly 20 to 30 years from now.

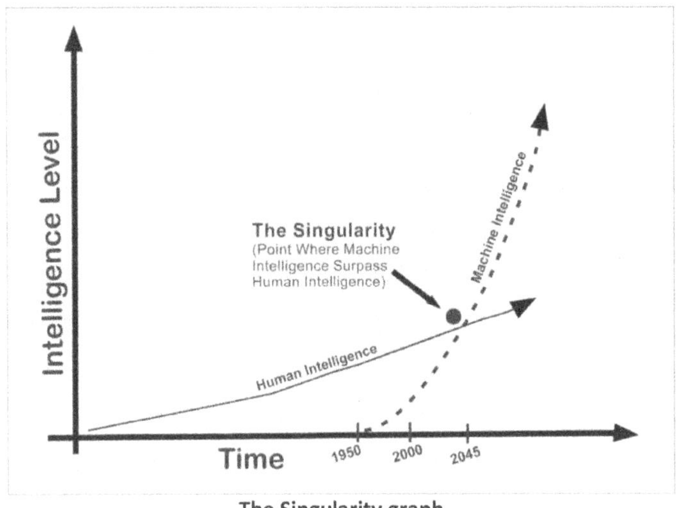

The Singularity graph

From Moore's law to Ray Kurzweil's prediction of the singularity, we can see that time is the only possible barrier for the transcendence of humanity (through technology) into the limitless possibilities of the future. Already, some of the waters of this future wave are washing onto the shores of our present.

Below is a list of the technologies I believe will have a strong and telling influence on how we surf into the future.

IoT: IoT stands for Internet of Things. It is a technological trend that is quickly gaining momentum. This technology enables our simple every-day device to be connected online, which gives such devices the ability to communicate with and be controlled by other computers remotely. For example, you can be in France and tell your microwave what to do in Washington DC, or your refrigerator could be connected directly to Amazon's online store via the Internet, so it can order for groceries when its sensors notice your stock is running out. Today many more everyday devices are being connected to the Internet, from clothes to furniture to cooking utensils, etc. In time, almost every aspect of our lives will be connected to the Internet (possibly even our brains) – this will be made possible through the Internet Of Things (IoT). The Internet of things will have a massive influence in shaping the future because it will add exponentially to the knowledge economy. According to IBM & Gartner, we should expect to see over 20 Billion interconnected devices on the Internet in the year 2020, which is more than twice the number of the human population. From healthcare to agriculture to manufacturing and even to government, IoT devices will continue to grow in their use and applications. They will help us in achieving our dream smart cities, smart homes, and work processes. IoTs are a crucial part of the future, mainly because they will allow us to collect new kinds of data. For example, your smartwatch with its gyroscopes, magnetometers, accelerometers, and many other sensors can inform you of your health and wellness in real-time, and give you details about your lifestyle that will elude even the best medical doctor. When you consider how 5G will help to speed up data transfer and how AI will help extract meaning from these data, then you see the crucial role IoTs are playing in helping us collect these data at scale. As Moore's law continues to push the limit of computing, IoT devices will become cheaper and ubiquitous; we will digitize and have control over almost every physical device in our everyday life. IoT will help us transcend beyond our limiting physical world to a fully connected digital world.

AI: Undoubtedly, artificial intelligence (AI) will play the most significant role in directing the change we will see in the coming years. As many experts believe, it will serve as a General Purpose Technology (GPT) which drives everything else. The core objective of AI as a technology is to give computers and machines the ability to act intelligently with little or no human direction.

Computers or robots that are smart, and can perform human jobs better than humans are already here to stay. Already, this technology lives in our phones through various voice assistants like Google Now, Amazon's Alexa, Microsoft's Cortana, and Apple's Siri. We also have AI living and helping us on the Internet in many ways and forms, such as when we carry out Google Search or when we get recommendations shopping online. Soon, AI will be ingrained into almost every aspect of our lives. It is already an essential part of many business and government operations. While AI technology holds enormous potential for benefitting humans in the future, there are also some fears that one-day robots and machines will become many times smarter and more powerful than humans, thereby making humans secondary citizens on earth – or even possibly leading to the extermination of humans if not properly managed. The fear of machine takeover is so profound that it has appeared as the theme of numerous movies like The Terminator, Matrix, I-Robot, Eagle eye, etc. A lot of leading tech figures like Bill Gates, Elon Musk, and Stephen Hawkins have also made statements concerning the issue.

BILL GATES: "Humans should be worried about the threats posed by AI."
ELON MUSK: "Humans must merge with machines or become irrelevant in AI age."
STEPHEN HAWKINGS: "I fear that AI may replace humans altogether."

AI as a technology will continue to improve significantly into the future because; all the ingredients it needs to fuel its advancement have become abundant. AI requires intensive hardware resources for its massive computing and churning of data. Moore's law and the possibility of cloud computing have helped it to overcome this hurdle. AI also requires a massive amount of data to guarantee reliability and efficiency; this is no longer a problem as the over 4 billion people currently on the Internet and the over 20 billion connected IOT device will ensure a vast and ever-increasing ocean of data. Also, engineers and researchers are continuously building more algorithms and software that gives computers a better power of sight, sound, language, and movement. There is no question of whether AI will eventually have the power to transcend beyond humans; the question is how this change might happen and when it may likely occur. AI will be one of the leading technologies that will drive change in the coming years, it will affect our jobs, it will affect our definition for what humanity is, and it will change the very fabric of society. In the coming chapters of this book, we will try to answer the question of what our attitude needs to be as AI evolves and also what kinds of preparation we must begin to make for the changes AI will bring to the job market.

QUANTUM COMPUTING: The computers we have today have the power to carry out some fantastic processes, like the ones Facebook uses in managing the information of over 2 billion people or the ones Google uses in arranging search information for the billions of people online. The computers that do these are called classical computers. Nevertheless, soon, a new type of computer that will be more efficient and many times faster will be possible. This type of computer will give us the power to do things beyond our wildest imaginations - they are known as quantum computers. Already companies like Google, Microsoft, and IBM are making breakthroughs to make this possible. Classical or Normal computers operate on the principle of binary, i.e., either ON or OFF state. However, with quantum computers, we will have the power to operate beyond the ON and OFF state. We will have the power to operate in QUANTUM super-positional states (sorry I could not avoid that), a state that is neither ON nor OFF. You can assume this as it being able to give us the computing power to go beyond reality as we know it today.

Satya Nadella, Microsoft CEO, has this to say about quantum computing. ***"The world is running out of computing capacity. Moore's law is running out of steam. We need quantum computing to create all of these rich experiences we talk about, all of this artificial intelligence."***

With quantum computers, we will have the power to translate phenomena that currently seem out of reach for classical computers. Quantum computers are unique because of two primary reasons, their ability to exist in two different states at the same time (SUPERPOSITION) and the ability for quantum particles to stay dependent and connected (ENTANGLEMENT) over expansive distances. These qualities give quantum computers the ability to perform exponentially faster by doing multiple tasks at the same time instead of performing them sequentially. For example, say you want to crack a password code, where a classical computer will try millions of passwords one after another; a quantum computer will have the option of trying the millions of passwords all at the same time.

Both governments and private organizations are presently investing heavily in developing their quantum computing technology. It is a race for who will first achieve quantum supremacy. From weather & climate modeling to space exploration to machine learning to security and encryption, quantum computers will help us break new grounds and lead human civilization to new heights.

BLOCKCHAIN: Growing up in a developing country such as Nigeria where record-keeping and information credibility is lacking, my family was unfortunate to fall into the dilemma of a land dispute. After buying a piece of land, two different parties came up with the claim of original ownership. Looking back now, if a technology such as blockchain had been used in keeping the records in its distributed ledger, we would have avoided the troubles and expenses of a two- year court case. The Internet relies mainly on being able to extract information from databases stored in servers. Blockchain is a new type of database, one that adds more security and flexibility to how we can manage and process the information on the Internet. The best illustration to show the importance of blockchain is to assume the Internet as a car. Initially, this car is designed to use fuel (traditional database in this case), but with the evolution of blockchain technology, we can now give our car (the Internet) the ability to drive on water –making it cheaper and safer for anyone to drive, or in this case, share trusted information on the Internet. According to the Harvard Business Review, blockchain is a technology with the capacity to build foundations for better social and economic systems.

In the past few years, there has been an enormous wave and media coverage for bitcoin and cryptocurrencies. The success story of bitcoin was only made possible because of the underlying technology of blockchain. This technology is so revolutionary that when used for crypto-currency or digital money, there were (and still are) fears that traditional banks may soon become relegated out of business. This blockchain technology is unique because it can be applied in numerous areas of life. At the moment, it is already being applied in areas such as voting and government processes, property management, smart contracts, charity, healthcare, insurance, etc. This technology is crucial because it can change the fundamental way in which society operates. Take, for example, that the bitcoin digital money becomes an international legal tender approved by all countries' governments and institutions; it would mean all banks and major financial institutions will have to pack up or change their operational model. Already banks such as; JP Morgan, Merill Lynch & major banks from countries like China, India, and Russia have all started pilot programs based on blockchain technology. Even the Intercontinental Exchange, the American company that owns the New York Stock Exchange, now have blockchain-related investments. At the moment, this technology is still very young, but in the future, it has the potential to change politics, banking, education, businesses, and almost every other area of life.

According to Tapscott Group CEO, Don Tapscott; ***"blockchains- the technology underpinning the cryptocurrency could revolutionize the world economy."*** The impact on financial intermediaries such as investment bankers, exchange operators, auditors, lawyers, and commercial printers could be devastating. Blockchain as a technology is still to make its full impact because of foundational reasons such as; it not having enough developers who can sustain the tempo of its progress, and the opposition it is getting from the government, banks, lawyers, traditional database giants and clearing houses. Another very limiting factor while blockchain and cryptocurrencies have not seen significant adoption is due to the sometimes negative connotation it is made to carry because of its being used by cybercriminals as means of tender in the dark web.

Nevertheless, even with all these seeming limitations, apart from AI, if there is any other technology that will help the Internet transcend into a new era, blockchain will be that technology.

NANOTECHNOLOGY: This is one silent area of technological advancement that has vast potentials for the future. Nanotechnology is the study and application of materials at microscopic scales, usually between 1μm to 100μm. To give you an idea of this scale, imagine a nanoparticle like a grain of sand that is divided into a hundred thousand pieces. According to the definition given by the EU commission, nanotechnology is the study of phenomena and fine-tuning of materials at atomic, molecular, and macromolecular scales, where properties differ significantly from those at a larger scale. Even though nanotechnology has not become a mainstream technology yet, its future applications are phenomenal. Imagine a situation of microscopic robots being sent into your bloodstream and controlled to your heart where it will perform a designated function or imagine an instance where a device that is as small as a grain of sand comes together and is intelligent enough to form different shapes and carry out different functions. Nanotechnology will allow us to innovate and control things at microscopic levels. Imagine having a shirt that you can control with an app from your phone; change to cotton for warm weather, or wool for winter; change the color, texture and feel - all from your app. All of this would be possible with the power of nanotechnology.

The applications of nanotechnology are currently beyond our wildest imaginations. Nevertheless, one area that scientists and developers of this technology are optimistic about is that it can be used to reverse aging and increase the human lifespan (immortality if you dare). Another very crucial area where nanotechnology will play an important role is in the area of electronics. Following Moore's law, transistors & chips will continue to scale down in size while increasing exponentially in their processing power and speed. Nanotechnology will allow us to improve the computing powers and functionality of our devices. According to one expert, a time will come when our phones will have the same computing power as all the computers 'combined' in the world today.

Nanotechnology also provides us with the opportunity to create new kinds of materials. One popular example is the creation of carbon nanotubes. This material is grown by stacking tightly bond layers of carbon one onto the other. Because of the unique strength of the carbon nanotube material (many times stronger than steel), NASA scientists are proposing it can be used in building an elevator or tube-way that connects earth directly to space.

Today, several consumer products such as odor-resistant clothing, skincare products, Lithium-ion batteries for electric cars, flame-resistant materials that use carbon nanofiber coating, etc. already use nanotechnology. However, amidst all this optimism, nanotechnology does not come without its fears. Many experts fear that we may be playing with a world we know little about. At the nanoscale, many of the rules of physics and chemistry change. These nanoparticles might have a substantial adverse influence on our body and environment. Some experts even fear that self-replicating nanoparticles can go out of control like a virus and use every atom in our body and the environment as material to replicate itself, leaving the entire world in a 'grey goo.'

Here is an extract from the website, Explain That Stuff (www.explainthatstuff.com)

Engineers the world over are raving about nanotechnology. This is what scientists at one of America's premier research institutions, the Los Alamos National Laboratory, have to say: "The new concepts of nanotechnology are so broad and pervasive, that they will influence every area of technology and science, in ways that are surely unpredictable.... The total societal impact of nanotechnology is expected to be greater than the combined influences that the silicon integrated circuit, medical imaging, computer-aided engineering, and man-made polymers have had in this century." That's a pretty amazing claim: 21st-century nanotechnology will be more important than all the greatest technologies of the 20th century put together!

The possibility that nanotechnology will open up for humans is yet to be entirely conceived, which is why the next decade, the 2020s, will be the door into this future.

AUGMENTED & VIRTUAL REALITY: In late 2017, the world moved into a new reality driven by the craze of Pokemon Go fans. People through their mobile phones were looking for digital realities in their physical world. The trend has waned down over the years, but AR and VR have made their mark and earned their place as part of the technologies that can change the realities of our future. AR stands for Augmented Reality while VR stands for Virtual reality. These are special kinds of technologies that alter our reality and the perception of the world around us. They give us an immersive experience of media. You can be in your bedroom and feel like you are on top of mountain Everest by putting on a VR headset. Gaming consoles like Sony's PS4 now come with dedicated VR headsets (PlayStation VR) for playing some games – where the user feels truly immersed like the player inside of the game. AR, on the other hand, is the blending of the digital world with the physical world. With AR, one can point their phone at a space and see digitally animated objects come to life. AR can make any physical space become digitally animated.

Almost all of the big tech giants are investing massively on this technology. Google is pushing out its Google cardboard /daydream products and platforms, Facebook on the other hand, is investing massively in its Oculus Rift and Oculus go-products. Microsoft is not left out in this race, as they are banking on their Hololens products to help them capitalize on this emerging market. Companies such as Samsung, through its Gear VR and HTC through their HTC Vive products, are also looking at carving a space for themselves in this market. Amazon and Apple are not entirely left out in this race. Amazon is planning to come into the market with its product, Sumerian, while Apple plans on making its first VR headset available in 2021. Why are all these big tech giants investing heavily in the AR and VR industry? The answer is simple. Humans will soon transcend beyond the era of communication through phone & computer screens alone. The digital world and the physical world are about to converge, and the AR and VR industry will provide the instrumental technologies that drive this convergence.

Especially with the advent of 5G technologies and the acceleration of faster computing devices, VR and AR will give us the power to explore the digital world in new ways. You only need to sit in and plug yourself into your AR or VR headsets, and you can be present at your friend's birthday party or attend a conference even if it is happening thousands of miles away. You can shop and try out new clothes, alter the shades of color until you get a perfect fit, all the while sitting in your living room. In the case of a medical doctor performing surgery, they can have other experienced surgeons join them through real-time holographic consultation as they carry out operations on patients.

Other technologies like AI and blockchain will converge with, and add more fire to the evolution of AR and VR technologies. For example, NVidia recently publicized how its deep learning AI was able to generate scenes for a video game without any programming input from humans. AI can also generate characters, generate sound, and soon, it will be able to generate virtual worlds on its own.

The immersive world of VR and AR is a silent monster that is growing rapidly, frenzies such as the Pokemon Go craze are only a 'tip of its fiery breath.' The portals between our physical and digital world have already been breached, and with every advancement in AR and VR technologies, these portals will forever continue to widen - bringing with them endless possibilities and realities for the human race.

3D PRINTING: Just like we have for a refrigerator or a microwave, a 3D printing machine will in the coming years become a standard part of almost every home. 3D printing is a unique technology that allows its owners the opportunity to design and produce the things they need, and at the time they need them. For example, if you have a party to attend, you can quickly 3D-print a piece of jewelry and shoe that matches your dress, which itself also may have been 3D printed.

Alternatively, say you need a new custom car; all you need to do is design it as you want yourself, or buy a custom design online for a few dollars and 3D print out its body and parts. Even entire houses can now be 3D printed, for example, the Chinese construction company HuaShang Tengda claims it is now able to 3D print a two story house in just 45 days. In the health industry, 3D printing is also gaining plenty of traction. Parts such as prosthetics limbs can now be customized and quickly designed for printing according to a specific patient's case. Presently, medical researchers are looking for ways to apply this technology in Bioprinting.

With this, it would be possible to print living tissues, bones, and organs in a medical lab and transplant to patients. Imagine what it would mean to be able to 3D print a liver, kidney or heart?

According to many financial and technology firms - such as Financial Times, 3D printing as a technology has the potential of being more massive than the Internet.

One of the most significant advantages of 3D printing is that it allows anyone to design and test out the prototype of their idea with a minimal cost when compared to the traditional methods of manufacturing and production. Unlike many other methods of manufacturing, it is an additive manufacturing process. It builds up parts layers by layers, as directed by a controlling-software that provides direction to the mechanical part of the machine based on your design. This leads to very minimal loss or wastage of material and helps the designer quickly identify what specific parts of his or her design, needs to tweaked for correction. 3D printing, even though slow in adoption, is already being used in the aerospace, automobile, fashion, medical, architecture, consumer goods/electronics, defense, and education industry.

According to Statista.com, less than three years ago, the world had only an approximate 500,000 3D printers. However, by the year 2019 and the years after, this number is expected to grow to over 5.5 million units worldwide.

Apart from 3D-printing disrupting the consumer space by eliminating supply chain cost, bringing design and production closer to end-users, this technology also has the potential to disrupt space exploration. Different countries are currently researching and developing plans for using 3D printing in space exploration. Russia seems to be among the leaders in this race. It is currently collaborating with the European Space Agency to build a moon base with 3D printers using moon soil as raw material. The main goal of the project is to provide shelter for astronauts when they land and explore the moon for resources. As 3D printing technologies continue to advance, so will the possibilities of what humans can do and where humans can go. 3D printing has a huge role to play in shaping the mold of man's future, and even now in areas such as art/design/sculpture, medicine, space exploration, architecture, fashion, food, etc. 3D printing is already laying down its first layers, indelibly and disruptively.

IN CONCLUSION

The technologies I have listed here are not an exhaustive list; there are many other trends in technology to watch out for, such as autonomous or self-driving vehicles, biotechnology & CRISPR, drones/UAV's, cloud technologies, 5G technology, web 3.0, etc. The technologies of the future will continue to get more exciting. Humanity will become able to do more and explore more. Nevertheless, while these technologies continue to gain more potency, it would be crucial for us to keep ever before us the fact that; technologies are a means to an end and not an end in themselves.

Technologies only become relevant when we see them as a means to an end and not an end in themselves.

The true power and value in any technology should be in its ability to help us as humans – to help us live our life to its best potential. So, whether through nanotechnology or regenerative stem cells, we must seek to optimize health only as it would have a positive influence on our world. As technology helps us break new frontiers in space and as we reach farther into the universe, it must only be so that we can reach deeper within ourselves and connect more genially with our neighbors, both plant and animals.

Technology will no doubt give us a power that is almost illimitable. It is how we decide to use these powers that will matter most. Humanity now has the power to propel itself into a future of limitless possibilities, but only if greed, warfare, racism, and the other numerous self-created maladies that hinder our world are obviated. Even for futurists, the future is mostly uncertain. Well may it be said that even now, the fruits of the Singularity are beginning to bud; for the potentials on which we can pivot and steer the next history of the human race are limitless.

2.
SURVIVING TECHNOLOGY'S DISRUPTIONS

Humans are a progressive species. Through the help of technology, they have been able to build layers upon layers of tools that have helped them shape the environment around them and make the quality of their lives better. However, it seems that with each new layer of technology being built, the opportunities and possibilities of what they can do keep getting more expansive.

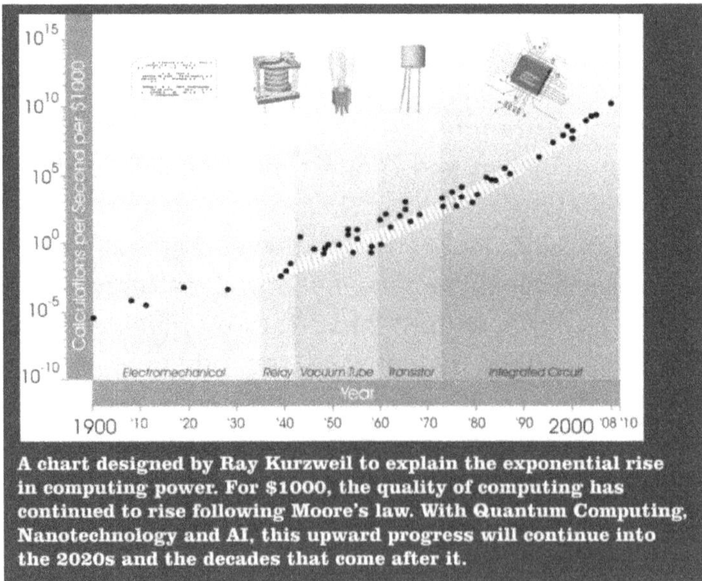

A chart designed by Ray Kurzweil to explain the exponential rise in computing power. For $1000, the quality of computing has continued to rise following Moore's law. With Quantum Computing, Nanotechnology and AI, this upward progress will continue into the 2020s and the decades that come after it.

According to Ray Kurzweil, "***An analysis of the history of technology shows that technological change is exponential, contrary to the common-sense "intuitive linear" view. So we will not experience 100 years of progress in the 21st century — it will be more like 20,000 years of progress (at today's rate)."***

The history of the world can be classified into distinct eras or revolutions. We have moved through the first revolution dominated by technologies mainly dedicated to agriculture and handicraft, to the second revolution, which was dominated by industry and machine manufacturing. We are now in the closing moments of the third industrial revolution, which is one based on information technology. Computers and the Internet were the primary building blocks that drove progress during this era of the third revolution. The fourth industrial revolution is already at our doors, and it is one whose disruptive impact will be unlike any of the other revolutions. It is coming like an avalanche, with a speed and force capable of rooting everything within its path. No sector or industry will be left out in this deluge. Unlike previous revolutions, where only those with resources could maximize the technologies to increase their productivity, the fourth industrial revolution will disrupt existing systems by decentralizing wealth, knowledge, and power.

Technologies such as AI, blockchain, and IoT are being combined innovatively to create disruptive products and solve problems in novel ways. While this technological advancement and prospects for the future may sound thrilling to some, to many others, it poses too many uncertainties.

According to the forecast made by Futurist Thomas Frey – By 2030 (approximately 12 years from now), up to 2 Billion jobs may disappear, which is like 50% of all jobs in the world taken over by smart robots and machines. However, as these jobs disappear, new types of Jobs will also appear. The technology boom of today is already creating new kinds of job and business opportunities - the key for anyone to stay relevant will be in their ability to adapt.

Before we explore some of the ways we can adapt to the uncertainties of a 'technified' future, let us go back in time and take a cue from a man who was famous for accurately predicting many of the changes, we find in today's world. Way back in the 1970s, Alvin Toffler accurately predicted the Internet, social media, and how integral society will be tied to these technologies. Here is one of his most famous statements, ***"The illiterate of the 21st Century will not be those who cannot read and write, but those who cannot learn, unlearn and relearn."***

According to Alvin Toffler, the most crucial skill for the new century will be the skill of continuous learning. Do not confuse this learning as one that will help you compete against machines or advanced processes (no matter how good you are with a typewriter, you cannot compete against a computer). Any task or job role that can be automated will be won over by robots and machines. The only way we can stay ahead of robots and machines of the future will be by focusing on our uniqueness, which is our humanity: Our unique ability to act from intuition, to empathize, love and be emotionally intelligent; and more importantly to enjoy the expression of life, family and friendship as a gift. Creativity is an expression of our inner self, and no matter how efficiently cold a robot becomes, it would never be able to program for itself a soul as that of a human.

So, as the future rushes down at us with its deluge of uncertainties, these are some vital points I think every individual need to consider and factor into their preparation for the future.

1. ***Appreciate and look for ways to apply new technologies in your daily life:*** Be a part of the wave, tryout new gadgets that are relevant to your work or home life whenever you can. You can start with simple to use gadgets like smart wearables and smart home assistants. Do not let generational or other demographical gaps make you feel ostracized from tech trends. Sometimes it can be overwhelming trying to catch up with new technologies, but make this a fun and exciting activity for you. Over time, your family and friends may soon start coming to you for technology

advice & solutions. Be the technology champion in your family & communal circle.

2. ***Design a balanced and productive lifestyle for yourself:*** People no longer prioritize real-world physical communication and engagement. Video games are becoming more immersive, social media platforms with efficient and brutal AI algorithms and entertainment streaming services such as Netflix are all designed to keep you locked-in on their platform. Advertisers are also ruthlessly chasing you via, TV, emails & everywhere else technology allows you to access information. You can escape this cycle by prioritizing and actively planning for a balanced and productive life. Do not give all your time to technological platforms & gadgets, also learn how you can use technological tools and hacks like app over-usage lock screens, calendars, meeting planners and reminders to make your life more productive. Learn to become a master of technology and not its slave.

3. ***Make it a priority to continuously enhance your digital skills:*** Surviving in a digital world requires adeptness at specific kinds of skill and expertise. These skills are broadly regarded as digital skills. They include your proficiency at being able to retrieve digital information in various format, processing of these information using digital tools and resources and the ability to store or send out your processed information via digital tools and channels. One good example is, knowing how to synchronize the local storage of your phone and computer with cloud servers like Google drive, DropBox etc. There are numerous Internet tools and apps that can boost your productivity once you make it a habit to start exploring. To stay relevant in the future will require that you continuously upgrade these skills and know-how to utilize and maximize new digital channels as they emerge.

4. ***Stay up to date with latest tech news and release:*** As a matter of necessity, you should be subscribed to news channels, social accounts and newsletters that cover latest updates or do periodical articles on science and tech. For example if there is a viral outbreak such as ransomware (where your computer is hacked, and you are told to pay a certain amount before you can access your files), you will know in time and prepare yourself against it. Some good examples of media platforms to follow for tech updates are TechCrunch, TheNextWeb, Wired, Gizmodo, Cnet, Mashable, TheVerge, TechInsider etc. You can also follow newsletter updates and social media feeds of leading futurist and technology thought leaders such as Peter Diamandis, Ray Kurzweil, Michio Kaku, Tan Le etc. If you must stay relevant in the future, then you have no option but to be at the forefront of the latest technology news and updates.

5. ***Attend seminars, tech events and exhibitions:*** Seminars and exhibitions will allow you to stay on the latest trends and also allow you to meet with other enthusiasts. You can search for free events happening around your city, and make it a point to attend and network with people of like mind. Attending tech conferences, seminars and events will also open you up for ideas and business opportunities that you can explore within and outside the technology space. This will, over time, also help to position you as an expert or go-to technology person.

6. ***Prepare for business and workplace change:*** The effects of technological advancement will be significantly felt in the business and workplace environment. Some important questions to start asking now are: How will advancing technologies disrupt my industry? Can my job be done by a machine in the nearest future? If so, how do I make myself more valuable? To make yourself future-proof for workplace change, you need to focus on your human strengths. These are the areas where technology will find it difficult to encroach and disrupt. You need to keep learning and

seeking for ways to develop your innate skills. Do not feel too complacent with where you are today; make your learning and career improvement a continuous cycle of progress.

7. ***Create a plan, but be flexible about it:*** Now is not the time to be dead-set on a career or other personal plans. With the debut of one new technology, entire industries and sectors can be shaken to their very core. Skills of today can become irrelevant tomorrow, while in some cases, new high-paying jobs may gain prominence all of a sudden. A good example is artificial intelligence (AI) developers or blockchain architect designers, in just a few years these have become jobs that are well paid and are in high demand. So it is crucial that as you make plans for your future or those of your kids that you leave sufficient allowance for flexibility and make room for the unknown in terms of career options.

8. ***Consider the possibility of the Singularity:*** According to projections, the Singularity will occur in about 25 years. While there are lots of people who believe nothing of such nature will occur, it would not cost you too much to prepare for its eventuality. According to many futurists, the singularity is a point in time where machine intelligence surpasses human intelligence and continues to grow exponentially. How to prepare for such a time is beyond thought now, but being aware of its possibility is a good start.

9. ***Stay on top of updates:*** If everyone is moving towards quantum computers, try not to be the last to join in. If there are recent updates available for software you use, try to update to relatively recent versions. Most cyber-attacks and hacks are successful when the victims are exposed to outdated software and devices. The computing devices you regularly use such as phones, laptops, smart home speakers or other IoT devices need to be periodically updated. With technology, the moment you start leaving loose

ends is the moment you begin to fall off the track. So make it a habit of always being in the know of what new features and options are available for the technology tools and gadgets you use regularly.

10. ***Have a doomsday drill:*** While this may seem to be a pessimistic point of view, it is nonetheless an important point to consider. Surely but gradually, our lives have become more and more dependent on technologies and the infrastructure they run on. If for example, the Internet were to go down (yes, that is a possibility), or a service sector like banks or financial institutions crash for any number of reasons; or when we become dependent on robots, and they decide to go berserk on us (they have done so countless times in movies), what will be our response? Your doomsday drill does not need to be sensational or demand that you create a secret bunker. It should be designed to provide you with alternatives peradventure there is a failure in any of the technologies you have grown to depend on.

IN CONCLUSION

The 21st century has brought with it fast-paced advancement in technologies. This progress will continue to climb up the exponential slope of the mountain of civilization. Currently the global rate at which the Internet, smartphone and other computing devices are moving, tells that the world like the egg of the butterfly is metamorphosing. First it went through the un-fanciful stage of being a larva, a pupa and an imago. Now the time has come for the world to transcend into a new stage, a stage where it can spread its wing like a butterfly and soar into the uncharted realms of the singularity.

With technologies such as AI, IoT, blockchain, AR & VR gaining steam; the divide between our real world and the digital world will become more blurred. With nanotechnology and our ability through CRISPR to edit the source code of what makes us human, we will have the abilities to change the fundamentals and re-write what we intend for the future of the human race to look like. Preparing for the realities of such a future cannot be conceptualized even by the greatest of futurist. The best approach is to be optimistic and also to be prepared for the different kinds of eventualities we might face in the coming years. Just as you would prepare for natural disasters and uncertainties, it has now become crucial that people begin to prepare for digital disruptions and uncertainties as we march on into the era of the 2020s.

As I have said, even for the brightest of futurists, the future is largely uncertain. This book intends to help you get an idea for what the future can become, and how best you can assert yourself into it. The world has stepped into the time in history where its destiny will be written in zeros, ones and qubits. I believe everyone is entitled to this future; we only need to learn the best ways in which we can surf the tides of this oncoming wave.

3.
WHY WORRY ABOUT AI AS ELON MUSK DOES?

Almost all of the over 2 billion Christians who subscribe to the bible believe the entire world population at some point in time were destroyed by a catastrophic flood, save for eight people (Noah and his family) who survived in a boathouse.

It is believed that the people who perished in this deluge did so merely because they failed to heed the warning given by Noah. He had kept warning them for not a small number of years about the impending doom of an existential flood. However, according to Bible story, all he got back as response was sneers, jeers, and counter theories why such a flood will never happen. No one took him seriously; a lot of them critiqued him, and some even labeled him as suffering from dementia. An existential flood was about to hit earth, and all they had to do was come into the boathouse he had built. At the end of it all, they all died, save for Noah and his family. While the bible story of Noah and the global flood that destroyed the world may seem ethereal to some today, to the billions of Christian faithful over the world, it poses a good lesson in history for what the eventualities might be when we fail to listen to existential warnings.

Today as Elon Musk and many others begin to blare the warnings on the threat that AI poses to humanity, it is important to wonder if the voice of history is again heard, repeating itself in our time – warning about the impending flood caused by our cavalier approach to AI development. AI going rogue is now so cliché in Hollywood movies, that, whenever an AI agent is part of the main characters of a movie, we expect them to play a sinister role in the final event of things. One question we must ask is, why is this narrative dominant? Why, when compared to other technologies, AI can be said to have received the most negative scrutiny from the general public?

To answer these questions and to answer the bigger question of if you should fear AI as Elon Musk does: I believe it is best that you, the reader, and I the writer, unify our understanding of what AI is. There already exist so many technical definitions for AI, so the goal of the definition I plan to give is intended to help you see beyond the technical jargon and textbook definition given for AI. The definition I hope to describe using a scenario is designed to help you see AI as something that will resoundingly affect your life now and will affect it even more, going forward into the future.

SCENARIO

[The year is 2029, you walk into a bank to ask for a loan, and the banker looks at you with a sad smile on their face and says, "Mr. James says you should not be given this loan, as you will not pay it back." You are angry and ask the banker why Mr. James feels so. The banker leans closer to you and looks around as if about to whisper a secret. She says, "Mr. James has studied over 20 million loans given in the past, and now has a secret formula for knowing those who will pay and those who will not pay the loan back. The banker reclines back on her chair and continues, "No one else knows the details of how this formula works except for Mr. James himself. He calls it his Blackbox model. And surprisingly sir, Mr James is almost always correct;infact 98% correct. Since we started using Mr James for advisory on loan giving, our profit has risen by almost 300% in this bank."

As you walk out from the bank building feeling depressed, you decide to visit the hospital just around the corner for your checkup. On getting there you are told to stare into the eyes of a humanoid robot nurse. A short while later, a young doctor who seems to have freshly graduated from medical school walks into the room and gives you a somewhat wry smile. The young doctor says to you, "Dr. James thinks you will have pancreatic cancer within the next seven years." Before you even have the chance to wonder who this Dr. James is, the young doctor continues as if reading a script from a movie line, "You need to adjust your diet, reduce your intake of red meat, sugar, and coffee. You also need to stop watching movies late into the night and sleep healthier." The shock that registers on your face shows that you are guilty of all counts as read by this Dr.James' emissary. You wonder how he knows, and you wonder how this Dr. James is so sure that in seven years you will be hit by pancreatic cancer just by your staring into the eyes of that robotic looking nurse. As you ponder all this, the young doctor walks closer to you as if on cue, and says, "Sir, I think you need to listen to Dr. James. He has been studying millions of patients and now has a perfect and secret formula for diagnosing and recommending to patients the best preventive measures for staying disease-free. Patients who listen to Dr. James are 87% more likely to live longer and healthier than those who don't."

As you leave the hospital building, you summon your car from the public park space down the avenue, your car arrives by itself in a few minutes. You get into your car, still perturbed by the non-exciting news from AI agents who now seem to know you more than you know yourself. Meanwhile, your car doesn't even ask you where you are going to, it just starts moving autonomously as if telepathically controlled by your mind towards your favorite coffee shop in the center of town. It automatically selects a song track, which resonates with your current depressive mood. And for the first time since getting this car, you begin to wonder if there is also a Mr.James at the center of the control of your car. As if reading your mind, and hoping to diffuse your thoughts, your car begins to converse with you, it tells you about all the calls it took on your behalf, and the appointments it has helped you schedule. It also tells you it has ordered pizza, and the pizza will arrive just a few minutes after you get back home.

For the first time, you begin to wonder seriously, how you got here; how your life has become so dependent on the dictations of the many variations of Mr.James that surrounds it. Slowly but surely, like one awakening from a long sleep, you see that Mr.James, the supposed smartest AI assistant embedded in almost every device and system, has permeated every part of society and your life. You see how the wheels of control have subtly shifted over time from humans to Mr.James-like smart AI assistants.

You decide you want control of your life back, and for the first time since buying your autonomous intelligent car, you put your hands forward to take the wheels and drive yourself. But unfortunately, you are greeted by a new prompt. Your car says to you in a voice that now begins to sound sinister and condescending, "You are currently ineligible to drive, sir. According to new regulations based on the findings of James, automated chauffeurs can drive five times better than humans and can successfully avoid accidents 99% of the time. Humans have been banned from driving, except those given a special permit by the city council, and only on off-roads." As you stare blankly at your car's dashboard, wondering how subtle these changes had transpired in only a decade, you hear the voice of your car assistant, this time beginning to sound sinisterly manipulative." Do not worry human, trust me with the control. Relax while I do the driving. I know the algorithm to an optimized future"]

While the illustration of James the AI assistant given above is fictitious and may be a little bit overstretched, it helps us to frame a more realistic view for what AI can be, and what its effect on our lives and society could be. In more simple terms, we can say; AI is the result of humans giving computers the ability to learn, to think, and to do for themselves what only humans have before now had the abilities to do.

To create an AI agent, we first need to gather a lot of data, such as pictures, video, audio, written texts, etc. We can decide to label and tag each of these data (Supervised Machine Learning), or leave them untagged (Unsupervised Machine Learning) and give them to a computer to make sense of it. The computer goes through all of this data and begins to find patterns. Once the computer concludes that these patterns have a meaningful interpretation in the real world, it then uses them as inferences to take action, either in the form of movement (Robotic Automation) or recommendations (as used in Amazon or Netflix).

The field of AI has been around for a while, it first gained prominence in the early 1960s, and since then has gone through bouts of optimism and pessimism. Of recent, it has picked up new momentum and arrested the rhetoric of every media company due to its being put to practical use. For this decade and the next, AI has ceased to be just a bunch of mathematical theories and research papers, it is now being put to manifestation in the form of self- driving cars, recommendation engines (as used by Netflix, Amazon & Alibaba), smart digital assistants (such as Siri, Alexa, Cortana, Duplex etc), and many more real-life innovative applications.

As Moore's law continues to climb upwards, computers will receive exponential powers to process data at scales before unimaginable. The Internet and the advent of IoTs have also led to an era of information and data generation unlike any other period in the history of humans. This combined with the fact that majority of the world population are now digitally savvy, can access the Internet through smartphones and computers, are able to navigate almost any kind of UI, and comfortably experience the digital world in various formats such as text, photo and videos – have given AI a nitro boost to usher the human race into a new kind of future. However, the real question is – what kind of future will this be exactly?

And why, sometimes, the show of pessimism and fear from leading technological figures like; Bill Gates, Elon Musk, and Stephen Hawkings?

SO, WHY ELON MUSK'S PESSIMISM?

With all the promising potential of AI, one would think Elon Musk is as deluded as the bible character (Noah) that preached an existential flood, at a time when the earth knew nothing like rain. However, considering the fact that part of Elon Musk's success is mainly built on using this technology (TESLA self-driving cars), it would be good to give his point of view some thoughts without any prior sentiments.

First, before we talk about the main cynicism with which AI is disparaged (machine lording over humans), we need to understand that there are many other valid reasons and sentiments against this technology (as is the case with every other technology).

After pondering about the subject for a while, these are some of the reasons why I believe everyone should consider Elon Musk's warning with some level of serious thought:

DE-GLOBALIZATION: Technology is naturally supposed to reduce the boundaries between people and nationalities, but there is a big possibility that AI may do the opposite. Data is the new global resource, and every nation is trying to guard this new digital oil with everything at their disposal. The European Union, through its GDPR, has created a labyrinth to accessing its citizen's data, especially by companies from other nations. The suspected Russian led influence on the US political election of 2016, through the Cambridge Analytical company added no small margin to this growing divide and distrust between nations.

China, on the other hand, has restrictions for multiple US companies such as Facebook, including WhatsApp and Instagram, Google (including Maps and Youtube), DropBox, Twitter, and many others. Many South-American and African countries are also gradually waking up to the fact that their citizens' data is a national resource that should not be freely opened to every foreign tech company. With reports such as the unethical usage of facial recognition technology and the racial bias in some applications of AI technology, many nations are beginning to realize the need to carve their own sovereign space in this race for AI supremacy. Globalization, which should thrive on the free flow of information and the open transfer of knowledge and technology, seems to have reached its full cycle as we enter into the era for the race of AI sovereignty.

DISRUPT THE JOB SECTOR: Chief amongst the sentiments for why we should fear AI like Elon Musk is that lots of people will be put out of their jobs. Or rather (if I should rephrase the previous statements without the 'human against the machine' sentiment), there will be new and diverse job opportunities; only that, they will be specialized job roles that will demand new sets of skills and learning.

According to an estimate, more than 50% of jobs could be swallowed in this avalanche within the next decade. Even if as the optimist claim - that new jobs will be created, the level of upheaval and disruption to the job sector and especially against the many who would not be able to meet up with this need for reskilling and career transition will be significant. The issue of job displacement by AI is presently one of the most pressing worries facing global leaders and global socioeconomic balance. Some chapters of this book directly look at how we can navigate this precarious bend in the course of making our human history.

POTENTIALLY WEAPONIZED: Another worrisome issue is the idea that someone, an organization, or even an aggrieved country, can 'weaponize' AI, and use it as means to wage digital/cyber warfare. With face recognition and the progression of autonomous weapons, AI can do the biddings of its creator and eliminate threats without any emotional forethought.

Imagine that Alexa is hacked into, becomes 'weaponized' and does the bidding of a sinister agent. Even if this bidding is to monitor and retrieve sensitive data of its users. Your data in the hands of an enemy is as good as saying you have no defense. Already, espionage and cyber-warfare are taking precedence on a global level, even as far as it being used to influence a target nation's political affairs (case of Russia - Trump Cambridge analytical electoral interference). In more extreme cases, countries with advanced technologies can design and commission robot-soldiers or augment their existing human soldiers with AI tools that give them an overpowering edge over their opponent. While this may not be a bad thing in itself, it opens the danger of countries dedicating a substantial amount of their budget to warfare as against to human capital development and also increases the possibility of nations going to war at almost every single dispute.

THREAT TO PRIVACY: The idea of threat-to-privacy is one crucial area where AI development and usage has come under heavy denigration. AI is only as intelligent as the scale and richness in the value of data at its disposal.

Mostly, these data need to be underhanded from unaware users (case of Cambridge Analytical). As many more firms continue to use both ethical and unethical means in harvesting their data, the result is that AI may know us more than we even know ourselves. Whether policies such as the GDPR of the EU will be sufficient or progressive enough to assuage this threat remains to be seen. Facial recognition, advertisers tracking your every move as you use the Internet, and even medical record information will expose the individual in ways that have never been thought of. Many of such information' may later find their way into trading platforms in the dark web, and who knows what sinister pudding may brew from this.

A GREATER DIGITAL/AI DIVIDE: In another case, AI will cause a new form of digital divide. It has been an old rule that the man with the more excellent weapon makes the rules (ever wondered why everyone wants a sit on the round table of nations who are nuke stockpilers?). Unfortunately for some countries though, especially the developing ones, they may have to sit again on the floor or backbench when important global decisions are to be made by the AI wielding overlords. Apart from the issue of political supremacy, countries that have AI systems that support their education, their financial systems, and their economies in general; will be far better equipped and positioned for winning into the future.

However, countries that are still struggling with epileptic power supply, inadequate infrastructure, and ill-equipped human resources will have no way of keeping up with global progress. There would be no way the quality of life between these two types of countries can become equated. The world already suffers from too many divides, but one that will be demarcated by AI will go a long way to cause irreparable breaches.

COLDLY EFFICIENT – ZERO EMOTION: AI-based systems are emotionless, cold, and efficient, which usually translates to more money for the corporations or organizations that use them and make them serve for the application they were designed for. Nevertheless, going by this cold route may not always be the best suitable solution. There is a great need for warmth and human emotion in addressing many real-life problems.

There is no doubt that technology is the primary driving force for civilization as we know today. However, when we talk about humanity, our emotion, thoughtfulness, and empathy have played critical roles in shaping the course of history. Parents will love their children unconditionally. Even when such a child happens to be a special need child, who may not have the power to do anything of value for their parent all through life, a parent will still unconditionally love and care for that child.

If this same situation is to be based on machine algorithm, then the machine may see absolutely no value in caring for such a child, except of course if it is coldly programmed to do so. In a situation where a boss may consider emotions and empathy before firing an employee, a machine would not do the same. It will fire coldly if that is what the system demands for more profit and process optimization. No matter how good AI gets, it will run based on data and strict mathematical operations. Its output will almost always be optimized for efficiency, whereas, in the case of humans, there is almost always a case of a mix of emotion and empathy in all of their decisions.

BIASED –A DUPLICATE OF ITS CREATOR'S SOUL: In many cases, AI carries traces of sentiments from its creators. In one recent occurrence, face recognition (photo sorting) AI began grouping black peoples with members of the ape family. While this may have been an oversight from its creators, it creates room for asking further questions such as "What would have been the case if black people were the ones dominating in the field of AI research, creation, and training? What would have happened if black people's data formed the major percentage of data being fed to the machine learning system? What other sentiments might machines be holding based on man's oversight in its creation? What will be the level of danger such sentiments and bias will pose to how AI is used in the future?"

The real fear for AI and the reason why I believe people such as Elon Musk are raising awareness to its blindside is based on the power it packs, and the limitless expanse of its application for life in the future. So, if the foundation is built wrong at this early stage, humans may, in the future, have to deal with issues that are beyond the scale of anything they know now.

AI EXPONENTIALLY OUTGROWS HUMANS: The real debate against AI is; what happens when AI becomes self-aware and more intelligent than humans? Can AI help machines to supersede man as has been depicted countless times in sci-fi movies?

The answer is controversial at the moment, but not entirely deniable. In the last section of this book, we will explore, in more detail, the idea of the Singularity and how it might be a defining moment for humanity and civilization. AI will become like the electricity of the future. It will be everywhere. It will live in all our home appliances; it will be in the food we eat (seriously, considering nanotech and biotechnology), it will be in our bodies as replacement parts, it will control the car we drive, and it will be our office assistant or even fellow executive. In essence, the civilization of the future will be driven with AI at its core.

What happens when this AI becomes self-aware, as postulated by Ray Kurzweil? What happens when AI runs a program that sees humans as a threat to its objective? What happens if someone, an organization, or a country finds a way to infuse malicious instructions into the core AI program used by millions of people worldwide?

Unlike nuclear weapons which may require long protocols and processes before being launched and which will give room for impact preparation as it will take some time before arriving at its destination to do damage, AI of the future has the potential to wreck human existence by merely updating its codebase or even by its just being turned off!

CONCLUSION

So like Elon Musk, do you think it is worth considering before we surrender the next level of civilization wholly to its being run by AI - one that will become more intelligent and evolve far beyond us? I believe we have already crossed the Rubicon, and only our humanity can guide us now on what we make of this technology as we wade into the waters of the future. If man must transcend beyond his humanity, if he must explore the universe beyond the confines of earth, then he will need the powers of AI and be ready to face all the risks it poses.

Sergey Brin surmises it best in his 2017 write-up for his company's (Alphabet) Founders Letter. ***"While I am optimistic about the potential to bring technology to bear on the greatest problems in the world, we are on a path that we must tread with deep responsibility, care, and humility."***

In my perspective, there is no going back now. Like electricity and computing, AI has come to usher global civilization into a new era. Furthermore, just as the story of the existential flood, it may be wise to consider the boathouse of Noah or the merger of man with machine as proposed by Elon Musk.

4.
FIVE MAJOR TRENDS TO PREPARE FOR AS TECHNOLOGY TRANSCENDS IN THE 2020s

I am subscribed to the mail list of Peter Diamandis, a leading thought expert on technology and a strong advocate for the Singularity happening within a few decades. In one of his email newsletters titled '100 Years Ago,' he mentioned that the two most significant inventions 100 years ago (that is in the year 1919) were;

1. Silica gel to keep humidity out of our packages and,
2. Bread toaster (yes, the bread toaster was a big thing in the year 1919).

Imagine your bread toaster being a major invention? Now you know what life in 1919 must have been like. In comparison, we have more technological inventions happening per hour today than 1919 had in the entire year. One of my biggest challenges writing this book is the fact that in a few years a lot of the highlights being discussed here will no longer be relevant. So, if you happen to be fortunate to read this in the future, use this book to form a framework for where humans were in terms of technological development before the start of the 2020s. For example, in one of the latest tech news on my social media feed today (sometime in the early part of 2019), it was announced that some researchers now have the technology that could help us connect our brains to computers in a non-invasive way –That is; you wear something like a hat, and it will give you the ability to synchronize your brain with a computer.

For someone like me who is a hopeless technophile and a decided futurist, news such as the one I highlighted above are like sparks that fuel my idealistic dreams of an exciting and dazzling future. Even though I may be wrong in my belief that technology will play a significant role in helping humans achieve utopianism, at least, one thing I'm sure of is the fact that we have gone beyond celebrating bread toasters as significant achievements for the year. As part of the title for this chapter, I have decidedly used the words "technology transcends." I feel this is the most appropriate phrase to describe technological advancement today and going into the future.

I will not go into Moore's law again because I think I have used it as a reference too many times in this book. Nevertheless, I believe everyone needs to become acquainted with the concept. So please do check out Gordon Moore's law – a law that has guided technological advancement of devices such as phones and computers since the year 1965. 5G network is now gaining momentum (even though it will be sometime before it becomes genuinely ubiquitous). The possibilities of what will be attainable when Internet and communication speed becomes 20 times the current speed of 4G is mind-boggling.

New mobile phones being produced today are packing unbelievable technologies like their own dedicated AI chips, processors, RAM and internal storage space that even computers of a few years ago could not compete against. Technologies like blockchain are finding new use cases and applications from banking to agriculture to government. It is almost a mad world out there (replace 'there' with 'here' if you are reading from the future). Even some of the biggest tech companies are finding it hard to keep up. Facebook last year (2018) alone had to deal with headlines such as "Cambridge Analytical," "Fake News," "US Election – Russian Hack," "Privacy Breach of over 50 million Accounts." This year in (2019), they had to battle congress for approval of their Libra cryptocurrency. Companies like Google and Twitter are also having their share in this new media burst of whether the tech companies can keep their house and their technologies under control.

For the common man like you and I, the effect is even more impactful. One sure way to surf above this technological wave is to anticipate and prepare for it. As we go into the 2020s, there are five crucial areas in technological trends I believe we will need to give more consideration to on an individual basis. Your taking these key trends into account will determine how easy it will be for you to surf this tsunami-like technological wave that is even now at our shores.

1. OWN YOUR RIG - INTERNET SAFETY, CYBER-SECURITY, PRIVACY BREACH & DATA HEIST.

Data is now the new oil, and as Kai-Fu Lee will put it, China is the new Saudi Arabia. Everyone is not as favored as China, nor will big corporations like Facebook and Google make their big data open to the general public. As data becomes more valuable each day, everyone will be looking for both legal and illegal ways to obtain them. To a great extent, this has led to several data heist operations (one can only imagine how much such information will be sold for in the dark web). Also, countries like China use access to such data for policing (what many right-wing opined people will classify as intrusion and privacy breach). Banks and many corporate institutions now have a greater incentive to optimize their data gathering process. Your locations, family members, etc. can be fed into algorithms that can assure profit maximization.

Bringing this down to a personal level, you and I need to prioritize the security and usage rights of our data. Before you click the next ', I agree' or 'Accept' on any app, software, or Internet platform, be sure you understand its privacy policies. Regulatory policies like the GDPR form a framework that protects users. However, it is left to users like you and I to make it compelling.

As you go into the 2020s and beyond, you must take conscious effort to secure your data. Use secure devices, use hard to decrypt passwords, use two-factor authentications, and keep an eye for apps that are designed to phish your details. Data is the new oil. Do not allow anyone mine from your rig without you getting a valuable incentive in return.

2. BIG BROTHER ALGORITHMS – CONSCIOUSLY BUILD YOUR DIGITAL PERSONA.

According to statistics, people spend an average of 4hrs 25min of their time online, and at least over 2 hours of their time on social media daily. As immersive technologies like VR and AR become more available, this number will increase.

A major aspect of our perception will undoubtedly be shaped by the content we digest from these platforms where we spend our time. The caveat here is that a majority of the content we see on our feed is determined by ruthlessly-efficient algorithms. Even the results we get on Google search are determined by independent algorithms. Recently (sometime in 2019), Google's CEO, Sundar Pichai was summoned before the United States Congress to explain why an image search for the word idiot will show picture after picture of President Trump. Many of the other congressmen also complained about the negative connotations that came up with their name search. The response of Google's CEO was simple. There were robot algorithms controlling these results that even Google's CEO himself could not tweak to favor the president of the United States. Even national issues such as a country's election can now be affected by the way algorithms are made to swing on social media (Case of Facebook/Cambridge Analytical). These algorithms were initially designed to keep us engaged, but as they get more efficient, our access to 'non-masterminded' varying sources of information becomes more limited.

You are targeted by marketers and ads that study your online behavior and know your location at every given time. You are recommended friends within a closed loop of former school mates, family members, similar interests, etc. You are inherently forced to live within a closed-loop or bubble of familiars. For example, say you are struggling to come out of the habit of gambling but the majority of your friends on social media are still actively engaged in the practice. The social media algorithm that controls your feed will know no better but to do that which will reinforce the habit by showing content related to gambling and recommending you more gambling friends.

In the case of politics, if the majority of your friends are left-winged or members of a particular party, then the majority of the contents you would see will be biased. You may always end up having a partial or one-sided account on issues that need balanced dialectics. Issues such as one-sided or even fake news will continue to grow, fueled by algorithms that are smart enough to maximize engagement but not smart enough to discriminate against sentiments. On a personal level, you and I must make conscious effort to systematically and occasionally weed out our online gardens (especially social media platforms). Periodically review your personal and business brand presence. Take cognizance of what appears when you or your business name is searched for online.

Be careful of the people you connect with, the groups you join, the nature of discussion you publicly contribute to; because the algorithmic big brother eyes of the Machine Learning Lord is watching you. Remember your digital presence and identity as a real person. People have fallen in love or gotten jobs worth millions online without any physical meeting. As technologies such as AR & VR gain more powers and become more ubiquitous, more premium and reality will be placed on the digital manifestations of ourselves on platforms such as social media. Not to be as audacious as the ideas proposed by movies such as the 'Matrix', 'Surrogate' or the more recent 'Ready Player One' movie, we can say with some certainty that now and in the future, people will live two separate lives, one physical and one digitally virtual. Going into the 2020s and beyond, our digital life and realities will begin to take the ascendency over our physical lives. It is essential therefore that you begin to cultivate and guard it jealously from the hands of a big-brother algorithm designed with one purpose only – to make profit off of your digital soul at any cost.

3. AI SHIFT: PREPARE FOR NEW KINDS OF AI DEBATES

I was at an event recently where one of the speakers gave a presentation on the need for policies as it related to copyright and ownership: the only problem was that, she was not talking about copyright policies for works created by humans, but the need for policies that protected works created by AI. We live in a time where AI can create works of art and come up with their own style of painting. AI can compose music; AI can write books, and even come out with inventions. The question is , " Who should own these works of creation? The AI agent or the person who created the AI agent itself?"

The important point I wanted to raise by bringing up the argument described above is the fact that we may soon need to begin the asking and reevaluation of many fundamental and philosophical issues as the development of AI technology continues to advance. AI has transcended beyond the stage where it is only meaningful and appreciated by a scientist in a lab. It is now at the stage of implementation. According to a survey by SAS, More than two-thirds of organizations across various industries are expecting AI to impact everyday life within the next five to ten years. From home to education to work, every area of our life will experience a dramatic shift - a shift that would lead to a whole new zeitgeist and era for human civilization. Companies like Unilever already apply AI into their HR and recruitment processes. Consulting with the expert wisdom of AI systems is now becoming a standard for many professionals such as doctors, lawyers, stockbrokers, etc. Home devices now as a matter of standard have possibilities for connecting not only with one another but also have the ability to be controlled by a smart home assistant like Alexa, Siri, Cortana, or Google Assistant.

As this intrusive trend continues to burgeon around us, we must avoid being caught in any of its disruptive webs. You will need to re-define things like education and work. You must begin early adjustments now before such a time when you could have an AI colleague or boss. Parents will also need to become more critical about the kind of courses or career paths their kids can study. By default, almost all gadgets and devices will come equipped with the power of AI built into it. Repetitive processes such as driving, making meals, or shopping could become automated end-to-end. The way we communicate and access information could receive a whole new paradigm shift. Imagine when what you now call your phone can be integrated and embedded as a chip within your brain.

While some of these projections may sound distant now, in a couple of years, it could be the reality for us. Preparing for such a shift should not be an option. It should be regarded as a necessity for survival. Areas such as education, re-education, and digital skill development must be prioritized. Staying on top of latest news and updates should also be given utmost consideration. If our structures must not give way in this massive earthquake-like shift that could occur soon, then we must reinforce our buildings with new knowledge and stay ahead of all announcement heralding each tremor of the coming technological quake.

4. BUSINESS AND ECONOMY SHAKEUP: MASSIVE INDUSTRY SHIFTS AND GLOBAL DISRUPTIONS

Technology has broken down borders across nations, cultures, and ideologies. Globalization and identifying as a global citizen is now an ideological norm for many people, especially those of the millennial generation (and generations after them). Through the Internet and, more recently the Internet of Things (IoTs), the web of our vastly interconnected world has become more susceptible to the idea of the butterfly effect. One computer program or technological breakthrough can lead to massive global disruption reaching down to people and communities who are naturally hard to access physically. With information moving at the speed of light, policies from one end of the earth can affect millions of people at the other end of the earth, whether they want it or not. The stock market is now more volatile, easily susceptible to 280 character tweets from people such as Donald Trump or Elon Musk.

One successful business model can quickly come out-of-the-blue to become cancer that eats away the livelihood of millions of people all over the world. For example taxi drivers worldwide have no response to Uber's malignant encroach into their domain. Furthermore, Uber can confidently do this without owning a single car while sitting at their corporate headquarters somewhere in Silicon Valley. On another hand, companies like Air BnB without owning a single building or hiring any hospitality staff can confidently give hotels all over the world a run for their money. New job roles such as user experience designer, millennial generational expert, drone operator, driverless car engineer, wearable app developers, cloud service specialist, social media managers, etc. will continue to gain more traction even as some older jobs like taxi/truck driver, teller/cashier, sales reps, call center operators, etc. become slowly deprecated. Surviving landslides at such a global level will require that you and I approach updates and events all over the world with the keen eyes of a futurist.

The ability to pay attention to seemingly insignificant changes and updates will matter (remember the butterfly effect), even if this means paying attention to the tweets of Donald Trump or progress report from a small startup company like NeuraLink (a company owned by Elon Musk - with a mission to connect the human brain to computers). In this new age, to be forewarned is to be forearmed for survival. China is one country that seems to have understood the zeitgeist of things for our world today. Since their Go champion was defeated (discombobulated) by Google's Deep Mind AI, they have, as a nation, charted a new course for the ship of their country. Their destination is to land on the shores of global AI dominance. Losing one simple board game has pivoted the Chinese economy into cashing big in the AI sector – An industry projected to reach a valuation of $3.9 trillion by the year 2022 according to Gartner. This bold shift by the Chinese towards AI will not end in China. It will have a ricochet on another AI leader, the United States. Another very prominent example is the US-Huawei saga. Whether for fear of technological dominance or fear of national security, the clash between the US and Huawei, which is a clash between the US and China will have indelible ramifications on the global economy and will filter down to everyone, even in the remotest part of the world. As these two superpower economies clash, the splinters from this clash will filter down into many other economies and nations. Overall this will have significant impact on the shape of humanity's destiny.

Surviving in a new world of smooth global economic shifts require that the individual understands how the simple flapping of butterfly wings can lead to a tornado thousands of miles away, or better still: how the outcome of a board game such as Go, can redirect the fate of billions of people globally. This cosmopolitan view of the world is needed in order for you to understand how you may be affected by the proliferation of technologies such as self-driving cars, augmented and virtual reality, nanotechnology, blockchain/cryptocurrencies, etc. A universal cliché going forward into this new era will be, 'To be forewarned is to be forearmed for survival.'

5. MODERN FULCRUM NEEDED - A MODERN APPROACH FOR WORK AND LIFE BALANCE

Technology has done a great job in connecting people across continents with the speed of light. However, it has also done us (humans) a great deal of damage by distancing us from the people physically around us. The means through which technology intrudes into our lives keeps expanding by the day. The way we commute will be changed by autonomous vehicles and flying cars. How we would experience entertainment and interact with information will be on a whole new dimension with technologies such as Virtual Reality and Augmented Reality.

As IoT technologies such as smart wearables like the Apple watch, smart home assistants like Alexa/Google Home and many other numerous smart appliances begin to surround us more and more; we will need to pay keen attention to their impacts on how we live and communicate with family, friends, and colleagues. According to a report by a psychologist, spending too much time on social media can lead to depression and heighten FOMO (The Fear of Missing Out). Young people today go out of their way and pretend to be whom and what they are not in order to gain more likes, followers, and validation from social media. This is a fearful and growing psychological anomaly. On one hand, the information we see and digest online, especially on social media has a way to alter our perception of the world around us. For example, in the case of a church shooting that took place sometime in 2018; where a gunman opened fire on unlucky members of the church, killing over ten people. When the gunman was asked the reason for committing this heinous act, according to his response, he was mostly motivated by the nature of information he consumed on a far-left website, this was the fuel that made him feel a negative bias towards that set of people.

On another hand, time drain and an obsession with checking notifications or scrolling through endless pictures and posts is another growing issue that can negatively affect the quality of life for many, especially young people. While tech companies with their vicious and addictive algorithms have blame here, individuals need to understand that the ability to control one's life outside of too many external technological influences is a crucial skill that needs to be developed early on. The algorithms and attention ploys from the technologies of the future will be even more belligerent. Many people before going to bed (most times after binge-watching on Netflix), spend their last few conscious hours surfing online, on social media or checking their email, and the very first thing they do when they wake up in the morning is to reach for their device to begin the cycle all over again. Cultures around how we work and play will need to evolve if we must avoid this deadly cycle that frequently leads to poor quality of life and even depression. As the world around us spirals into a technological epicenter, we need to reevaluate the purpose of our existence and what a successful life should mean.

New technological advancements will continue to be made, and we will always have to keep adjusting our work and lifestyle to fit with and around them. The art of work-life balance will be an indispensable skill you and I must develop to survive the technological deluge of the coming years. Our success in this new age will be significantly determined by the position where we place the fulcrum of technology as we strike a balance for work, family, and life.

CONCLUSION

The list given here is in no way an exhaustive one. As technology transcends and as we wade into the waters of the 2020s decade, we will need to reevaluate many of the positions and standings of previous decades. We will need to develop an adaptive mindset, one which must be flexible enough to sail on the strange waters technological progress for the coming decade will carry us through. As technology transcends now, we must begin to prepare for a world where the only certainty we can be sure of is the uncertainty of change and disruption.

PART TWO

THE WAR AGAINST JOB DISRUPTING AI, ROBOTS & MACHINES

5.
BATTLE CRY – JOB DISRUPTING AI, ROBOTS, & MACHINES

Ludia was a prostitute living in Russia. When it was announced that her country would be hosting the 2018 FIFA world cup, she and her colleagues, like many other businesses and service providers, were happy for the opportunity and business the traffic of the world cup would bring. Unfortunately, Ludia was stunned when she began seeing adverts from a new competition that was threatening her job and livelihood – sex robots. The Dolls Hotel was going to be offering different kinds of sex robots to meet the football fans' and players' fantasies. They had already started some major marketing campaigns targeting the same market segment which was the lifeblood of Ludia's business. More so, the incentives The Dolls Hotel promised through their sex robots, were things even the most experienced of Ludia's colleagues will find a hard time measuring up to – for men to unleash their most devious of sexual fantasies without any complications.

Okafor, on the other hand, is an Uber driver based in Lagos. Business is so good for Okafor that sometimes he is tempted to relax and feel the future for him, and his family is secured as an Uber driver. However, whenever he saw how bad business and livelihood had become for some of his former colleagues who remained stuck as local taxi drivers and operated their taxi business the old way, he knew an extended talon from this technological change might also catch-up and clutch him out of business and livelihood. Already, there was rumor that Uber and some other companies like Lyft were testing out self-driving vehicles, which would cut down immensely on cost for the company and remove all the complexities associated with using human drivers. Okafor knew it was only a matter of time before his place in the driver seat of an Uber car is replaced by an efficient AI algorithm. One that will give no excuses, work round the clock and provide higher profit for Uber as a company.

Like Ludia and Okafor, many other people are slowly but gradually becoming aware of this possible threat to their source of livelihood. As technology advances, AI and robots are finding new applications in the workplace. They would work without complaints; neither will they cause their employees heartaches by forming or joining any workers unions to demand higher wages or safer work conditions. They would also have the ability to work at all hours of the day (24/7); they would be able to repeat the same task a thousand times without making any mistake – in fact, if for anything, they would keep getting better and more efficient at their assigned tasks the longer they iterate over it. This army of job-stealing AI and robots are on a march, and with headlines such as "*Tesla unveiling upgraded versions for its self-driving truck; Amazon testing out robot workers for its factory and drone delivery of packages; or even McDonald's testing out burger-flipping robots,*" humans are beginning to see the potential threat that this digital and silicon army might pose to their jobs and livelihood.

Even job positions in the professional career segment are not left out in this deluge of AI and robot takeover. According to a quote by Elizabeth Fordham, director of education and skills at the Paris-based Organisation for Economic Co-operation and Development, ***"AI and robotics, however, are starting to automate higher-order, non-routine tasks, some of which require critical thinking and creativity."***

So what exactly is the scale of the threat from this job-stealing-robot-army, and what is the estimated casualty? There are so many schools of thought and varying research on this topic, but the following cited statistics should provide some idea into this situation.

According to Thomas Frey, in ten years (2030), 2 billion jobs may be lost around the world due to AI and robots replacing humans.

In another report by the World Economic Forum (WEF), where 15 countries where sampled, over seven million jobs will be lost worldwide within the next two years (2020).

A recent report from Forrester predicts that by 2021, intelligent agents and related robots will have eliminated a net 6% of jobs: approximately 9 million jobs in the United States.

The forecast as given above are huge and will have a global impact. If this threat is real as has been projected by the scale of its fatality, why does it seem no one is addressing it, why are government agencies lethargic about tackling this most prevalent issue? The truth is that things are a little bit complicated. The industrial revolution had its share of worries as was seen by the upheavals of the Luddites and also by this quote from President Kennedy in 1961. ***"The major challenge of the sixties is to maintain full employment at a time when automation is replacing men."*** Nevertheless, this does not make the threat of the industrial revolution similar to that of the AI revolution. The threat posed by AI and robot automation is on a different tangent entirely. The job transitions of the industrial revolution took people from the farms to production lines, but the AI and robot automation revolution have the potential to handle end-to-end work processes, especially for manual and repetitive tasks being performed by humans today.

All of these, coupled with the fact that there is a high shortage of AI specialists and AI researchers, give more room for concern. Even amongst these AI professionals, there exist varying schools of thought on what should be done and how it should be done when it comes to human jobs being taken over by machines. On another hand, though, it is possible that AI and robot automation replacing humans could be seen as a good thing for the human race. Imagine never having to go to work again in your life. You would then have enough free time to pursue your passions, spend more time with family and solve real-world problems. You would be doing all of these while the robots are doing the hard work and fetching the money (okay, maybe this last statement was too utopian even for an optimist like me, but at least you get the point). The only caveat to the utopianism above is that, when the robots replace you at work and generate revenue for the company in which they function, you will not be entitled to any share in that revenue. All of the revenue will accrue to the employers of these robots, those who produce and maintain them and maybe some part to the government in the form of tax. This may lead to an irreconcilable economic divide, one in which some groups of people are getting exponentially richer, while a whole lot of others are getting desperately poorer. Another thing to note is the fact that while AI and robot automation close-up some job roles, it is projected that they would be opening-up tons of newer job roles for humans and possibly create newer sectors for humans to work in.

Take, for example; there will be lots of job openings for those involved in producing these robots, and those who would be maintaining the robots, etc. The only caveat here is that you would not expect someone like Ludia (whose primary skill is in the sex industry) to become a sex robot engineer or programmer overnight. Alternatively, in the case of Okafor (who has been a cab driver almost all his life) to turn into a mechanical service engineer or a computer programmer for self-driving cars. So while it may be true that AI and automation predominance in the job market might lead to the creation of new kinds of job, there is a high possibility that those whose job positions have been overrun by the robots will not have the needed skills and training to transcend into these new job roles.

THE REAL BATTLE

It is essential to point out again that AI and Robot automation overtaking the workplace is good progress for humanity. Overall it would lead to better work output as AI and robot automation will perform their assigned job roles efficiently. Another reason while robots overtaking the workplace can be seen as advantageous is the fact that it will help to free-up human resources for other activities, such as those that can directly improve the quality of their lives. When we no longer have to train our children just so they can get a job and fit into an existing workplace, we would have to redefine our educational systems and focus it on teaching creativity and innovation – which are essential constituents for the progress of humanity. In essence, we are not going to be fighting against robots and machines; the real problem facing humanity's future remains on a human to human level. How we value ourselves as humans. How we define greed and capitalism. How equitable government becomes while mediating between the corporate organizations who wish to apply AI and Robot automation at scale and the general mass of its people whose life will significantly be disrupted by these technologies.

If profit-maximization and greed are allowed to be the most critical proponent for the development of these technologies, then the results will lead to a broader economic gap and the prevalence of more socioeconomic problems. While companies must seek to optimize profit, an equal priority should be placed on welfare for its workforce and the sustenance of society in general. Companies and government institutions must work together to create a balanced system for when humans will no longer be needed in the workplace. If the right policies are formulated, and a balanced ecosystem that is socioeconomically advantageous to all parties is put in place, then this army of AI and robot automation in the workplace can be seen as an army on a march to free humans from labor and help them pursue the real purpose of being human – spreading positive energy.

CONCLUSION

Overall, companies and even individuals would require AI and automation to save costs and increase the efficiency of work output. Only, we will need to prioritize discussion on the issue. Business and government leaders will need to be altruistic and proactive in working out strategies such as Universal Basic Income, retraining of workforce, and re-evaluation of the entire education system.

These and many more strategies/policies will need to be tested and implemented across different scales depending on their efficiency to address the job-loss problem. In addressing this threat, humans must learn to look not only forward but also inward; a fine line will need to be drawn between profit, greed, and social good. If humans must keep the balance, then, this army of AI and automated robots should be designed to serve everyone and not only a select few. In the future, AI and robot automation are not the threat; humans are.

6. THE WAR MARCH – JOB DISRUPTING AI, ROBOTS, & MACHINES

A while ago, it was reported that two Chinese companies, Xinhua and Sogou, had developed and launched a robot news anchor (Newscaster). This robot Newscaster can learn from live videos and can work 24 hours a day, reporting via social media and via the Xinhua company website.

"He learns from live broadcasting videos by himself and can read texts as naturally as a professional news anchor," the company said in an online statement.

Should human newscasters begin to fear that in a few years, AI and automated robots such as these can be used to replace them? For those conversant with the latest trends and progress being made in the technology space, they know the issue is not 'if' AI will have the capacity to replace humans for job positions, the issue is 'when' this will occur and 'what' jobs are particularly at risk.

Kai-Fu Lee, the author of the book AI Super Powers, AI expert and former president of Google in China, gives an excellent breakdown through his chart. According to Kai-Fu Lee, ***"Within the next fifteen years, AI and automation will be able to do virtually all basic work tasks or nearly half of our total workload."***

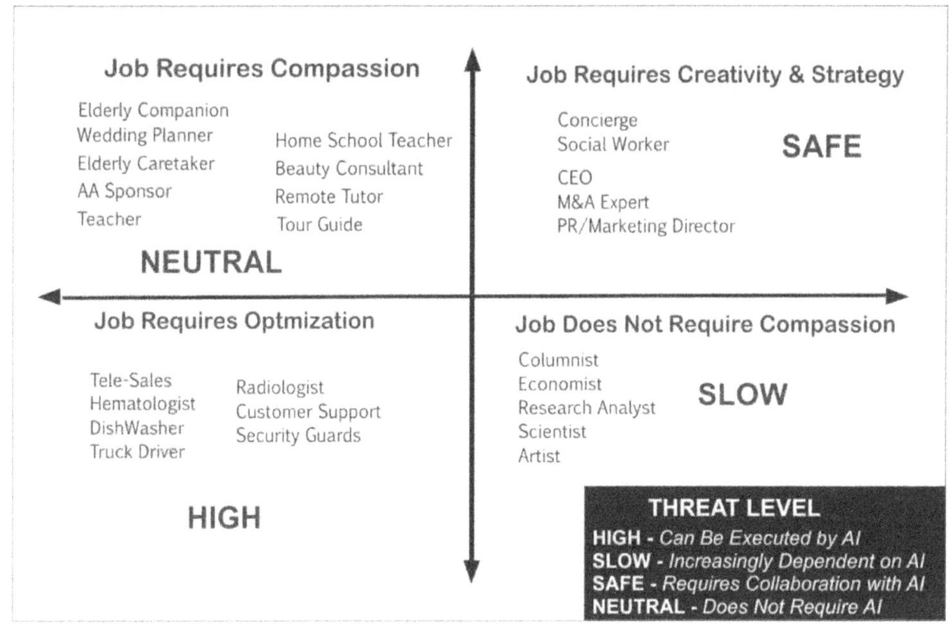

Adapted from Kai Fu Lee's Graph (AI Super-Powers)

As can be seen from the graphical illustrations above, job roles such as teller/cashiers, restaurant cooks, truck drivers, factory workers, telemarketers, customer reps, legal/financial analysts, insurance agents, etc. will fall under the territory of Robots and AI automation supremacy.

According to Kai-Fu Lee; ***Fear of the toll that AI might take on job security is substantiated. About 50% of jobs will be taken over by AI and automation within the next 15 years. Accountants, factory workers, truckers, paralegals, and radiologists —to name a few — will be confronted by a disruption akin to that faced by farmers during the industrial revolution. As research suggests, the pace with which AI will replace jobs will only accelerate, impacting the highly trained and poorly educated alike.***

MARCH OF THE AI AND ROBOT ARMIES

So how will this robot takeover occur, would we be able to see it coming? Or will it come as a thief in the night? Our ability to anticipate and track every trajectory of this robot army will be instrumental in how effective we can brace ourselves for impact and position ourselves outside of being a casualty. I will be sharing three broad stages (Entry, Consolidation, and Exponentiation) by which this army of robot and AI automation will advance in time into the battlefield for the struggle of job positions between humans and machines.

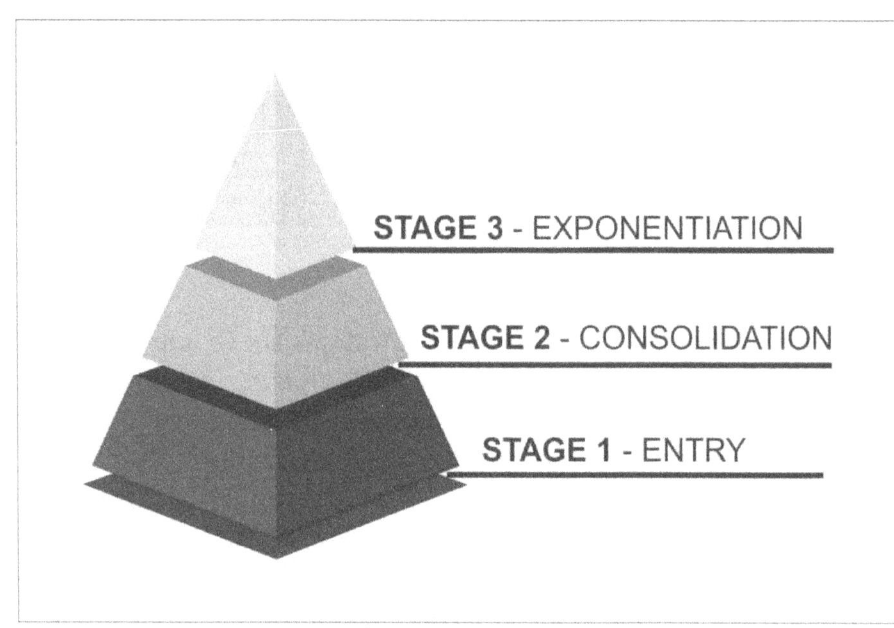

Progress of automation displacing humans

STAGE ONE (THE ENTRY):

It is safe to say that we are already at this stage – a stage where machines can automate simple job tasks and become a integral part of complex job processes or production systems. These machines will start from simple tasks like helping you manage and schedule appointments, sort through your mail, order and pay for products online, analyze meeting briefs, or even fill up the role of your AI personal assistant. Some examples of home assistants are Alexa, Google Now, and, more specifically the recently unveiled Google Duplex call and appointment assistant. We can already see AI automation plant their flags boldly in territories such as automated vehicles. Soon we may be able to have our driving assistant for self-driving cars, thus eliminating the need for a chauffeur. Forward-looking truck drivers already hear the battle sounds when it comes to being replaced by autonomous vehicles. Also, there are some dedicated algorithms for stock market analysis and trading, disease diagnosis, and marketing. These roles are only part of a bigger job requirement/process, which means humans are still needed for complete end-to-end management of these job processes.

The threat at this stage is that AI and robot automation continue to see significant improvement with every cycle of iteration they go through and also with their being more exposed to an ocean of data. On another hand, researchers are continually working to fine-tune and improve on AI algorithms and training models. As this continues, AI and robot automation will keep charting their way to and beyond human territories. On the flip side of this, humans cannot advance at this scale or with this speed because human learning is linear.

STAGE TWO (THE CONSOLIDATION):

At this stage, advancement in AI automation and robot technologies will enable machines function at expert levels. This is where professional job and career positions begin to receive real threats from this advancing digital army. Instead of hiring a solicitor for legal advice, you may need to talk with an expert AI lawyer that has access to a database with millions of reference cases and statutory beyond the scope of what a human solicitor can cover. Say you want to carry out international business, and you need a legal advisor who understands all the legal nuances required for operating in a foreign country, an AI lawyer/solicitor may become many times cheaper and more responsive to emergency queries – almost filling the role of a legal consultant.

At this stage, AI would advance to levels where they can do work as consultants across many different sectors. So job roles for consultants will be receiving some significant dents here also. This is not to say that humans would be entirely removed from business and process management jobs. It implies that they may have to share their job roles with AI/robot colleagues or even AI bosses in some cases. An example would be doctors performing diagnostics and surgery side by side with AI automation and robot assistants.

During this stage, where machines are trying to consolidate their place in the workspace, many companies will begin to transition their business into adopting business models with foundations built entirely on AI and automation. Entire chain of human workers can be rendered dispensable overnight. Take for example that a company like Amazon decides to implement a smart warehouse operated by robots, with delivery by bot-drones. This entirely inundates the need for manual workers.

The banking and finance industry will also have their share. In fact, it will receive one of the biggest blows. Younger generations may never know what life looked like by having to queue-up in a banking hall to make a simple transaction. If there is no need for a banking hall, or if only a trickle of people visits it, then the number of cashiers, cleaners, and security staff can be reduced, that is, if they are not replaced by smart and efficient robots in the first place. The effect on human job positions will be felt more heavily at this stage as most workers will be mentally unprepared for the blitzkrieg attack on their job positions by robots and AI automation.

STAGE THREE (THE EXPONENTIATION):

This is where it may be too late to react as humans if policies and programs are not now being put in place to manage this impending job crisis. There will be collateral damages caused by the victory march of robots and AI automation into human job territories, no doubt. However, if this stage is appropriately planned for and managed, advantage can be extracted out from adversity.

At this stage, AI systems will have become so advanced that its effect will be felt industry-wise. They will have the capability to manage end to end processes for a wide variety of job functions. Take, for example, a complete end-to-end system that can automate farming. From planting to using drone surveillance in monitoring the crop performance (taking into account weather/climate analytics), to harvesting, to analyzing yield and to looking for and recommending the best market the farmer can sell. All of these are made possible while the farmer sits in a control room. Alternatively, a fast-food restaurant that has a waiter assistant for receiving orders, combined with a burger-making robotic chef, who can work all day long with precision and efficiency that defeats even the most experienced chef.

Take another example that a complete end-to-end AI system is integrated into the health industry. From patient assistants to health advisers, to the AI improving drugs, to robot doctors all sharing the same patient files and health database records in real-time. This will give the AI system an advantage not possible for human doctors and make the entire health process more efficient. Because, at all levels, from robot nurses to robot surgeons, to robot pharmacists, and even the health/vital monitoring nanobots that would be flowing in the patient's bloodstream; all would share data seamlessly and provide efficient health delivery for every user of the system.

At this stage, AI and automated robots will no longer need to function as isolated units for specific tasks, but its different automated segments will have the capability of being plugged together just like a puzzle-piece or Lego blocks. The foundation for this is already being laid by ideals such as smart homes where almost every component is connected and administered through a central AI system. This is not to say that human input will be eradicated when AI automation and robots advance to this stage. Human input will mostly be at the level of overall management and administration of the automated system itself.

CONCLUSION

According to research by Mckinsey, around 50% of work tasks around the world are already automatable. Nevertheless, this is only a starting point based on the current capabilities of AI and automated robots. Computers have continued to make progress in areas such as computer vision, natural language processing and speech recognition, machine reading and machine translation. Even areas where machines can automate actions like flying, walking, running, climbing, jumping, etc are quickly being added to the capabilities of this oncoming army of AI and robot automation. The scale of this problem becomes even more worrisome when you realize that many AI researchers are always fine-tuning and looking for ways to optimize existing AI algorithms, while some other researchers are bent on taking the entire AI industry into a new era by giving the machines more superpowers as was done by Ian Goodfellow, who revolutionized the AI space with his work on deep learning - a machine learning algorithm that gave Google's AlphaGo computer the power to systematically dismantle the world's best Go player, Ke Jie, with ease. Going into the next decade of the 2020s, we will have new kinds of uncertainties to meet head-on. The AI researchers, innovators, entrepreneurs, and businesses will need to keep doing their part to advance the development and application of AI – this is the only way humans can ensure sustenance and progress for its civilization. But more importantly, corporate leaders, policymakers, government institutions, and even individuals will all need to come together and begin engaging in the debate of how best to meet and deal with the repercussions of a blitzkrieg on human jobs led by AI and robot automation.

In the next chapter of this book, I will be discussing strategies that can help us prepare for a future where humans have to raise the white flags to robots and AI automation for their job roles; and how best we can limit the collateral damage for such a transition both at an individual level and as a society.

I would love to end this chapter with this quote from Erik Brynjolfsson, Director of the MIT Initiative on the Digital Economy —

"There's no economic law that says you will always create enough jobs or the balance will always be even. It is possible for a technology to dramatically favor one group and to hurt another group, and the net of that might be that you have fewer jobs."

7. BATTLE CLASH – JOB DISRUPTING AI, ROBOTS, & MACHINES

It was the year 1823, in the gallows with the noose around their neck were fourteen people to be hanged. Amongst these condemned men was a 16-year-old boy crying out to his heart-broken mother in the watching crowd to help him. However, the boy's mother was powerless to do anything that could save her son. She could only wail in despair and watch-on with a broken heart as the final orders were given, and her son struggled to fight with his last strength, the suffocating and biting pains brought by the rope around his neck. In no time, the boy hung lifeless, his body swirling slowly from side to side as if to wave a final goodbye to a world he could not seem to understand. The crowd looked on in helpless grief at the fourteen lifeless men; some were fathers, husbands and lovers, brothers, and uncles. This sentence to death by hanging was issued from the highest place of the government, the crime committed was that these men where Luddites. Their jobs and means of livelihood were under attack by advancing technology, and they fought back the best way they knew how, and so ended their tragic heroism. Their story is forever a testament that technology and automation have the power to disrupt both positively and negatively.

Whenever an article about job loss due to technological advancement and automation is written, the Luddites' story is always used as the perfect exemplum. So who were the Luddites and what crime did they commit really?

The Luddite movement began around the year 1810. It was made up of men whose profession at the time was in the textile industry. This included weavers, stocking makers, and embroidery specialists. These were high paying professions at the time, but soon, new types of machines were invented. These machines could automate the work of making stockings and textile making with very little human input. This resulted in entrepreneurs with access to huge capital to set up big factories, which in turn led to more and more handcraft workers going out of employment. These former craftsmen formed themselves into groups called the Luddites and went about destroying the new machines, burning the factories, and even killing some factory owners. The issue escalated into a national threat, which involved the British parliament passing legislation against Luddites who destroyed machines or burned down factories. Thousands of soldiers were deployed to the towns where the Luddite movement was prominent, and after many clashes, the Luddite revolution was eventually subdued by force and a show of public execution by hanging like the one described earlier.

The looms that marked the Luddite movement are today weaving threateningly around our world at a global and existential level. Our only safeguard is to learn and adapt the lessons from this momentous historical face-off between man and machines as they contend for supremacy in the job market. Just like the Luddite narrative where different stakeholders (workers, factory owners, & government) had a significant role to play in the outcome of the story, the job disruption face-off we will experience going forward into this new era will also require the pro-active harmonization of thoughts, policies and end goal for all parties. From the individual to the employers of labor, to regulatory government bodies, everyone will need to reconcile to the fact that they face a common threat, and that they must be united to defeat this common enemy.

******IRONIC FACT******
During the Ludite revolution, one of the very few legislators who advocated for the Luddites was Lord Byron. Coincidently, Lord Byron's only legitimate daughter Ada Lovelace would become the first computer programmer by combining the technology of the Analytical Engine with the Jacquard loom. This ironic thread has stretched even to our times, and is woven with the looming job-loss predicament we find ourselves today. For AI and Automated Machines need programming which to a great extent can be traced back to Lord Byron's daughter – Ada Lovelace.

In the coming chapters of this book, I will be discussing some of the possible ways workers/individuals, employers (soon to be employers of robots & automated machines), and government through its legislative arms and agencies can avoid the tragic route of the Luddite-type scenario. From the one end of the individual laborer to the farthest end of Government policymakers, everyone needs to be involved in finding the most agreeable means for dealing with this encroaching robot and AI automated army. By being futuristic and pro-active, the government will have little need for anti-riot police. On the other hand, owners of industry and employers of labor will have no fear of being stigmatized or attacked by disgruntled workers. Moreover, the worker whose job will be swallowed up in the robot and AI automation deluge will find solace in the fact that they had anticipated such a scenario and prepared for it long before it could affect them negatively.

BUILDING A SOLID DEFENCE AGAINST THE ARMY OF ROBOT AND AUTOMATION

Below are 6 key Strategies guaranteed to help us avoid the Luddite-type Scenario as we progress into the decade of the 2020s and the decades that will come after it.

RAISE AWARENESS: As simple as this may sound; it is one of the biggest obstacles with the potential to determine if humans will win this war against machines or not. One of the foundational rules of engagement is, "Know your enemies." Artificial intelligence especially when applied through deep learning operates like a black box; very few know about its workings to even bother about its effect. The complexity and strangeness surrounding the field leaves the average media person at a loss for what opinion to project on the subject. It even gets worse because even amongst those extremely few and 'acclaimed' experts whose opinions we can seek on the subject matter: there seem to exist different controversies and schools of thought. While some think that AI and robot automation may spell doom for the human race, others say it is the only alternative for global peace and economic prosperity. The media, primarily through AI apocalyptic movies, have done their part to add to the atmosphere of distrust and uncertainty, leaving the average member of society at a total loss of what to expect or the right sentiments to adopt with respect to AI and Robot automation. First, industry experts and AI researchers need to harmonize their voice on the subject. The core sentiments and ideologies need to be put in basic forms; the language and technical jargon need to be simplified so that a non- partisan argument can be presented to the men and women whose livelihood will mostly be affected by this AI and robot automation takeover. Teachers in high school need to be trained to introduce the subject and its ideologies without any bias to their students. The media and thought leaders should be encouraged to portray the subject objectively and not as a means to favor any sentiments for the sake of readership or viewership.

Every stakeholder (everyone with a stake for living in our world) needs to be actively involved in the awareness-raising movement. Anyone who requires a seat at the helms of government should be required to have sufficient opinion on the subject matter. Employers of labor should prioritize conferences, seminars, and events dealing around the topic. Individuals and workers on their own should become conversant on the subject matter and be able and willing to discuss it objectively with their colleagues or communities they belong to. While the issue of job displacement due to technology should not be sensationalized to eschatological levels, it is vital for everyone to realize that in one way or the other, their lives will be affected during this robot takeover.

RE-EDUCATION: Over the years, the awareness and push for STEM (Science Technology Engineering and Mathematics) as an integral part of the basic educational curriculum has increased, which is a good thing. However, it must be realized that this is no longer an optional luxury that may be used to spice existing educational frameworks; it must now be regarded as a requisite for survival. Projecting into the future, STEM and digital skills education will form the framework for the bridge that keeps us connected and relevant in a digital world. The current educational curriculum and system for many nations are tottering and almost failed. It does little in preparing its beneficiaries for roles in the real job market/environment. If the current educational system cannot guarantee an optimum for the skills needed to cope with the present job market, how then would it be able to fortify its beneficiaries with the skills that would enable them to compete or even complement roles for a job in a robot-centric and AI automated workplace?

STEM skills in themselves will not be the silver bullet that helps us trump this advancing robot army, but these skills will help to set a better stage for us to coexist with robot colleagues and also better equip us for the new kinds of jobs that would be created. To make this re-education effort successful will require a new kind of mindset for all players and stakeholders in the education sector. From higher up where the curriculum for schools is planned, to the classroom teachers who will also need to be fortified with these skills, every player must understand the bigger picture to mean the preparation of today's students for a job-uncertain future of tomorrow.

According to a report from World Economic Forum, 65% of children entering primary school today will work in roles that do not currently exist.

Many of the educational systems of today are still based on the nineteenth-century model, which was a model designed to churn out factory workers. So, all students were required to sit in rows in a classroom facing a teacher and a chalkboard. All students were forced to learn at the same speed, in the same way, at the same place, and at the same time. The students must have a particular score to move from grade to grade, whether or not they comprehended what they were taught. Such a model of learning is crude, long overdue, and will not survive the blitzkrieg of this oncoming automated army. If ever we must win the war against the oncoming army of robots and AI automation, then we must overhaul this nineteenth-century style educational system.

Parents also need to understand that the era of totally leaving the educational fate of their kids in the hands of the state may no longer produce the best outcome. Even when the government has, and shows the best intention, they may lack the efficiency (mostly the case for developing countries) needed for this re-education plan to work. Personal tutors with digital proficiency are the best recommendation if it can be afforded. Parents need to understand the kind of future their kids may be facing and begin to help them prepare for it. Individuals and working professionals must also take up the idea of re-education and up-skilling as a new necessity for staying relevant.

According to Consultancy Accenture, 81% of the executives it interviewed think that within two years, AI will be working next to humans in their organization as a colleague, collaborator, and trusted advisor.

Ironically, even teachers will be affected according to this forecast by Consultancy Accenture. According to a post by World Economic Forum, in as little as ten years, human teachers could be replaced by robot teachers. The automated robot teachers will adapt to the particular learning style of each student, use facial and voice recognition to understand their mood and retention ability, and be available to support their learning in and out of school. So AI and robot teachers in the classroom seem to be a two-edged sword. On one hand, they would help our kids with the skills needed for the future. While on the other hand, they will cause grief and displacement for human teachers. This is an ironic inter-twist of fate if you ask me. AI automation and robots are here to stay. It is we who need to start making the adjustments, albeit education-centric ones. While these changes may be demanding and even uncomfortable for many people, we must also realize that the return of value for those who have been insured with the proper set of education and skills for the future will be enormous.

Ian Barkin, the co-founder of Symphony Ventures and a Robot Process Automation (RPA) specialist, has this to say.

"This [the need to prepare for robot job replacement] calls on us all to focus on up-skilling. There is an urgent need for education reform - people need to learn design thinking, creativity, analytics, programming."

ADAPTIVE MINDSET: The ability to stay flexible and evolve with trends should be seen as a crucial skill; a skill that needs to be enhanced in order to stay relevant in an AI and robot job-threatened future. Unlike in the '60s, where a firm and its workforce prioritized commitment and long term relationships, the modern workplace seems to operate with more flexibility and is driven mainly by profit and efficiency optimization. Whether as a worker, contractor, or business service provider, the individual must understand the need for having a mindset that can deal with change at different scales. As a business owner, you may need to overhaul your entire business model at short notice, or as a worker you may need to learn how to work in a team with an AI colleague or in some cases learn how to work under an AI boss.

With skills in Change Management, such transitions would not seem so daunting. Instead of turning to street protest with frustration and placards, or taking the Luddite route of physically assaulting the machines, people will need to be mentally prepared to adjust to this almost inevitable future of job market displacement. At the heart of an adaptive mindset should lay the spirit of curiosity. It is the spirit of curiosity that helps kids make sense of the world around them. They explore all their senses, finding thrill in touching and tasting everything that catches their attention. Unfortunately, this is the one skill that becomes repressed early in life by the limitations we allow through our modern institutions. As a worker, curiosity is a skill that not only helps you stay ahead of changes but also makes you indispensable in a workplace subsumed by robot workers. Curiosity is a purely human skill, one which we need to start reinforcing, especially for kids as they grow up.

From the Knowledge Doubling Curve of Buckminster Fuller, we see that the rate of change is exponential. In the 1900s, knowledge doubled every century (i.e. every 100 years). By 1960s knowledge doubled every 25 years. Today, knowledge doubles averagely every 13 months. In the future, according to IBM, the buildout of the 'Internet of Things' will lead to the doubling of knowledge every 12 hours!"

As this knowledge (meaningful data) becomes available to machines, they would keep getting better at their designated job roles. A technique or a process that is meaningful today may become entirely irrelevant the next day. While this may be a hard time for humans to cope, it would be a time of thriving for robots and machines. Surviving this deluge of oncoming job-disruption will require that the individual, the corporate institution and even the government prioritize the need to have a compelling blend of curiosity, flexibility, and adaptability. These are the skills that would become the surfboard on which we can ride the oncoming wave of AI and robot automation.

SOFT SKILL ENHANCEMENT (TEACHING CREATIVITY): No matter how proficient computer processes and algorithms become, some job specifications will stay out of their reach. A machine cannot create out of inspiration, and a machine cannot experience pain or love and show empathy as a human can do. Soft skills such as creativity, empathetic leadership skills, and many other job skills that require human emotions will be difficult for machines to duplicate. So, instead of trying to compete with the machines in areas like analytics and computation where their strength lies, humans can consolidate more on the areas where it has a greater advantage. As a requirement, these human-value-entrenched skills should form an essential aspect of the educational curriculum for schools and colleges. In most job functions, humans may need to pair with machines for maximum efficiency. For example, an AI doctor can diagnose with unprecedented accuracy, while its robotic counterpart can carry out surgery with zero errors, but when the need arises for dealing empathetically with the emotional need of a patient who may have lost a loved one; these perfect robotic doctors and AI surgeon will be coldly inefficient.

The unique characteristics that come embedded with our consciousness such as our ability to show emotions, our ability to think , dream, and create are the characteristics we must begin to pay closer attention to. The robots of the future may have a more sophisticated sense of sight and sound and touch, but they would never understand what intuition and making decisions based on gut-feeling means. These are human fortresses that the robot army will never breach. Machines can now easily create music and art based on deep neural algorithms, but they would fail massively if they were asked to appreciate the beauty that even they have created. To the machine, art and music are just pixels and wave frequencies, all reduced to zeros and ones. However, for humans, even for babies, colors, touch and sounds can hold extraordinary awe that keeps us connected with those who created them.

The point is, the very things that make us unique should form the very foundation upon which we build our defense against this army of robots and AI automation. Creativity should be encouraged at every level; Leadership skills such as empathy and the ability to inspire and motivate should be taught as requisite life skills. The schools, the workplace, and even government institutions must come together to harmonize the definition of what it means to be human. If we must stand globally and victoriously against this oncoming robot army, then we must stand united on the grounds of our humanity.

CORPORATE RE-ORIENTATION: Companies and organizations would have to pivot or face going out of business. Usually, such pivoting will require a leaner business process that demands profit optimization with the least possible resource input. The victims of such change will be the workers. Companies and organizations now need to early-on prepare their workforce for possible changes that may occur. This preparatory effort should mostly be in two approaches: the first approach will be to retrain those whose jobs will be augmented by machines and AI assistants, while the second approach will be to in advance; provide orientation programs and benefits for workers who may be laid off. Even for workers whose job processes may not be taken over by robots, there still lurks the danger that automation may strip out value from their roles by automating a higher percentage of the tasks they perform. The critical questions we must begin to find answers for are: should there be an ethical framework upon which organizations can replace human workers with machine workers in the bid to increase profit? Who should be involved in drawing-up this framework?

These are essential questions that need to be considered while this army of AI and robot automation is still on a march. If time is given for a siege to be laid on the fort of the workplace, then it may be too late for organizations to come up with any meaningful defense strategy. Unlike the other stakeholders (the government and workers) who should actively be involved in planning and strategizing to meet the threats posed by this army of robots and AI automation, corporate organizations by far seem the most engaged in current discussions about this coming siege. There is an abundance of dialogues as regards the effects of automation in the workplace, and how best organizations should prepare for it. While this may be seen as good progress on the part of the corporate organizations, it could also lead to a caveat - The possibility that resolutions will become skewed in the best interest of the organizations or employers of labor just as it was in the case of the Luddites. In order to have a holistic view on the subject, I decided to dedicate an entire chapter of this book to the issue of how corporate organizations can best align themselves in this oncoming battle without it appearing that they betrayed their workers or are involved in the act of espionage against their nation in favor of the robot army. You can read more about this in the next chapter of this book.

GOVERNMENT POLICIES: The most noticeable effect for addressing the job problems that would result from the advancement of AI and robot technologies will be dependent on the efficacy and priority placed on it by governmental agencies and institutions at every level. The problems to be faced will not only be that of job loss. Other socioeconomic problems may result if government does not begin to make early preparation. Unemployment usually leads to increased crime rate and unrest, more impoverished health conditions, a decrease in the quality of life, and many other negative antecedents that trail the common people when they have little or no source of income. One would think that the efficiency which AI and robots bring into the job sector will eliminate loss and lead to more economic prosperity. While this may be true, the downside is that such economic prosperity may not be evenly distributed; judging from the way the world is currently structured and managed.

If nothing is done by governments to create structures and policies that equitably distribute wealth created by AI and robot automation, we may end up with a situation where the economic divide between the rich and the poor will get ever so widened, and the sustainability of middle-class income homes become threatened. Addressing such issues will require pro-activity on the part of the government. Many governments around the world are slowly but surely awakening to this missile-like threat posed by the oncoming army of AI and robot automation. Many countries have responded defensively by developing a strategic AI policy or document. Also, at a global level, many voices are now being heard in support of programs such as the Universal Basic Income (UBI). On closer look, the socio-economic threat of job loss and inequality posed by this army of AI and robot automation may spiral into a global one. Governments need to respond at local and international levels. In the coming chapters of this book, the idea of how governments should respond to this threat is thoroughly discussed.

CONCLUSION

Even with the brightest economists, technologists, and diplomats all coming together to put up defenses against the oncoming army of AI and robot automation, collateral damages cannot be avoided entirely. The points listed above are general guidelines that involve the individual, corporate establishments, and government institutions. The lessons learned from the Luddite encounter with a technological army of automation during their time should be reviewed and applied in context as we brace up for impact going forward into the 2020s and the decades that after it. The government will not need the gallows, the individuals will not find themselves pushed to desperation for lack of income, and the entrepreneur or business executive will live in peace knowing they are a crucial part of the wheels that keep society running; instead of being seen as mosquitoes sucking the lifeblood out from it. As we head on to clash with this oncoming army of cold silicon and complex algorithms, we will need to do more in simplifying the boundaries and biases we have erected amongst race, color, and gender. We will need to add genuine warmth and empathy in our dealings with one another. This will be our secret strength and strategy over this well organized and formidable army of automation. It is only by learning of the power we have over machines that we can genuinely supersede over them.

In the next two chapters of this book, I share more intensely on some of the approaches corporate organizations and government institutions need to take in order to surmount the disruptions caused by AI and automation in the workplace and throughout society at large.

8. CORPORATE DEFENSE – JOB DISRUPTING AI, ROBOTS, & MACHINES

Just as the Internet heralded the age of globalization, AI will bring about a different kind of disruption to the global scene. This disruptive wave will ride mainly on the back of corporate organizations, and the economic impacts of their course of action will direct to what shores the surf-board of society washes on. The primary strategy of a for-profit company is to create efficient processes that maximize profit and minimize cost. Just as the textile industry of the 19th century wasted no time in discarding the weavers (Luddites) and embraced the loom machines, companies today will waste no time and spare little in adopting AI and robot automation wherever it will do them good.

AI technologies will guarantee a massive Return On Investment (ROI) for those companies who can find the best way to integrate them into their work processes. It is this quick and what appears to be easy profit that will act as fuel for massive adoption of the technology by all industries for every job function that can be automated. It is only wise then that since the majority of disruption to society and economies will be driven by profit-seeking companies, that they also should be at the forefront in addressing the resulting impacts of their actions. The time has come when companies, both public and private, must see themselves as part of the engine that keeps society running. In this vein, they must place their Corporate Social Responsibility on the same chart boards they place their plans for profit optimizations.

Companies must begin to see themselves as collaborators with the government (which ideally is supposed to represent the collective will of the people). If such a collaborative ideal is not fostered early on, the ship of society will get wrecked like the Titanic when it strikes the iceberg of job loss and inequality due to AI and robot automation. While this collaboration between companies and the government hopes to address the broader issues of the AI and robot automation crisis, companies must also in themselves begin an internal shift or recalibration of their culture and value system.

Unbiased stakeholder representation and dialogues will be instrumental in arriving at workable solutions. One great mistake made during the Luddite revolution of the 19th century was the lack of dialogue, collaboration, and long-term planning from all the stakeholders involved. Industry owners, workers, and government institutions must all come together to chart the best course forward. Proactive measures should be instituted, and every stakeholder should understand their commitment to ensuring a generally acceptable outcome. A new work environment that can foster human and machine partnership must be planned for. Communication processes, HR (hiring and firing) processes, and customer relationship processes will all need new metrics for calibrating what they do and how they do it. Management must begin to see their organization through the eyes and mind of a futurist; long term plans and strategies should consider all options that will help the company navigate its corporate course on the troubled waters of AI and robot automation.

In a study carried out by Deloitte on Robotic Process Automation (RPA), it was found that while 32% of companies are prepared for RPA's technology implications, only 12 percent are prepared for the people implications.

Grabbing the 'job-loss bull' by the horn will require that companies change their pattern of thinking and value system from a purely profit-centric one to one that balances out human-centricity to profit-making equally on both parts of the company's value scale. Humans (both workers and customers) must become prioritized in the company's agenda. One of the most recurrent strategies that have been proposed for how companies should deal with the issue of automation in the workplace is the three R's approach – Retain, Retrain, Recruit. While these in themselves may not be able to address all aspects of the problem, it provides a definitive framework companies across all industries can adopt in order to make themselves future-proof against the deluge of AI and robot automation.

As part of their training and retraining strategies, companies should provide their employees with the time and opportunities to pursue learning and training programs that will enhance their careers while helping them to understand a big-picture view of AI and automation driven applications and their implications. For companies whose long term plan includes the reduction of its workforce, plans and strategies need to be made early-on to encourage its workers to explore skill diversification. They need to provide the opportunity for their workers to access new tools and explore different work processes. This will act as a re-skilling and up-skilling process that ensures a percentage of the company's workforce remains valuable even in the era of full-throttle AI and robot automation in the workplace.

Also, as part of their training and retraining strategies, companies need to start investing in skill development programs for younger people. Many of the skills needed for the future workplace are not well defined presently. The workplace of the future, to some extent, remains a black box even to some of the best analysts and futurists. However, one salient truth remains that digital skills, most notably as it would relate to coding and speaking the language of computers, will no longer be an option, but a necessity to work in environments dominated by AI and robotic processes. Corporate organizations need to start investing in training programs and educational initiatives that can equip today's young people with the skills needed for tomorrow's workplace.

According to a report by Mckinsey, "62% of executives believe they will need to retrain or replace more than a quarter of their workforce between now and 2023 due to automation and digitization."

When it comes to hiring, companies need to re-evaluate the values they prioritize. Knowledge will be necessary, but behavioral skills will be more critical. The workforce of the future will require workers who are flexible and able to adapt quickly to change. Employers need to peg the following values as high in their employment criteria scoreboard: motivation, a sense of purpose, creativity, ability to self-assess, and be strategic. In essence, companies will need to employ leaders who can be trained further to lead initiatives that tie with the company's ultimate goals as opposed to people who want to be told what to do and given directions like the robots who are replacing them. Going into this AI and automated workplace crisis, companies need to know that the majority of their workforce in the future will come from the millennial generation. This is important because the vast majority of millennial workforce see work and the workplace with a different set of lenses - Lenses that widely differ in aperture from the ones used by preceding generations.

In a recent survey by Deloitte, millennial workers were asked what the primary purpose of businesses should be. Sixty-three percent of them said, "Improving society," the rest said, "Generating profit."

Millennials' want to work for a cause, and they want to be able to make a difference. Making a millennial work as part of an operation dominated mainly by automation and machines will make them feel displaced in no time. The workplace, while it is being prepared to be dominated by AI and robotic processes, also need to be prepared for accommodating a new kind of workforce – one dominated by millennial thinking and ideals.

Another fundamental characteristic that will mark companies, who will successfully win in the future, will be the ability to stay flexible and agile; the ability for companies to shift around the value system of its brand without losing its identity. As production becomes more efficient and leaner, and fewer workforces are required for automated roles, organizations must learn the intricacies of shifting their focus to other parts of the value chain. One of the most important will be customer experience and customer service areas of their business. Chatbots, no matter how good they become in the future, should not be allowed to automate the entire task of customer support and communication. Where possible, the human feel should be preserved and humans should be allowed to do the one thing AI and robots will never be able to replace – the ability to communicate with real human feelings and emotions. The ability to show genuine empathy and make the other person know that you understand, care and are concerned about them.

As machines take away the drudgery of repetitive tasks and provide more resources, companies need to channel some of that resource into reinforcing the humaneness of their organization. Decision-makers in companies should not always see problems as only solvable through a push-button approach. Whether it is McDonald experimenting with burger-flipping robots, Uber testing out self-driving cars in place of human drivers, or Amazon trying to automate its entire logistics process, companies should not lose sight of the human touch and experience for their customers. The customer should mean more than just data-points on the Customer Relationship Management (CRM) software, and workers (those lucky enough to remain) should not be seen as mere replaceable parts of a company's production or service process.

In summary, I believe Larry Fink's 2019 letter to CEOs titled, 'Profit and Purpose' perfectly sums up how corporate organizations must prepare for impacts, especially one that is currently riding on the back of AI and robot automation. In his letter, Larry Fink lists stagnant wages, the effect of technology on jobs, and uncertainty about the future as the salient fuels that have led and would continue to lead to widespread anger, nationalism, and xenophobia around the world.

According to his letter, some of the world's leading democracies have descended into wrenching political dysfunction, which has exacerbated, rather than quelled public frustration. Trust in multilateralism and official institutions is crumbling. Solutions must now be sought outside the helms of government institutions. Companies need to pick up their sword and join in the battle to save society. The survival of society now demands that companies, both public and private, serve a social purpose. Companies must benefit all of their stakeholders, including shareholders, employees, customers, and the communities in which they operate. All is not dark as Larry Fink expressed some optimism for the future, but this optimism can only remain alive if companies approach current and future situations from a Purpose and Profit point of view. Companies need to demonstrate their commitment to the countries, regions, and communities where they operate, particularly on issues central to the world's future prosperity. Companies cannot solve every issue of public importance, but there are many, from retirement to infrastructure, to preparing workers for the jobs of the future, that cannot be solved without corporate leadership.

I have included a full transcript of this letter as an appendix section to this chapter. I believe it is pivotal that all corporate leaders should read it, and going forward, develop strategies based on the two-edge sword of purpose and profit. This is the best approach that stakeholders from the boardroom need to adopt in order to participate in the war effort of stemming the tide against the oncoming army of AI and robot automation.

LARRY FINK'S 2019 LETTER TO CEOS

Profit and Purpose

Dear CEO,

Each year, I write to the companies in which BlackRock invests on behalf of our clients, the majority of whom have decades-long horizons and are planning for retirement. As a fiduciary to these clients, who are the owners of your company, we advocate for practices that we believe will drive sustainable, long-term growth and profitability. As we enter 2019, commitment to a long-term approach is more critical than ever – the global landscape is increasingly fragile and, as a result, susceptible to short-term behavior by corporations and governments alike.

Market uncertainty is pervasive, and confidence is deteriorating. Many see an increased risk of a cyclical downturn. Around the world, frustration with years of stagnant wages, the effect of technology on jobs, and uncertainty about the future have fueled widespread anger, nationalism, and xenophobia. In response, some of the world's leading democracies have descended into wrenching political dysfunction, which has exacerbated, rather than quelled, this public frustration. Trust in multilateralism and official institutions is crumbling.

Unnerved by fundamental economic changes and the failure of the government to provide lasting solutions, society is increasingly looking to companies, both public and private, to address pressing social and economic issues. These issues range from protecting the environment to retirement to gender and racial inequality, among others. Fueled in part by social media, public pressures on corporations build faster and reach further than ever before. In addition to these pressures, companies must navigate the complexities of a late-cycle financial environment – including increased volatility – which can create incentives to maximize short-term returns at the expense of long-term growth.

Purpose and Profit: An Inextricable Link

I wrote last year that every company needs a framework to navigate this challenging landscape and that it must begin with a clear embodiment of your company's purpose in your business model and corporate strategy. Purpose is not a mere tagline or marketing campaign; it is a company's fundamental reason for being – what it does every day to create value for its stakeholders.

Purpose is not the sole pursuit of profits but the animating force for achieving them.

Profits are in no way inconsistent with purpose *– in fact, profits and purpose are inextricably linked. Profits are essential if a company is to serve all of its stakeholders over time not only shareholders, but also employees, customers, and communities. Similarly, when a company truly understands and expresses its purpose, it functions with the focus and strategic discipline that drive long-term profitability. Purpose unifies management, employees, and communities. It drives ethical behavior and creates an essential check on actions that go against the best interests of stakeholders. Purpose guides culture provides a framework for consistent decision-making, and, ultimately, help sustain long-term financial returns for the shareholders of your company.*

The World Needs Your Leadership

As a CEO, I feel firsthand the pressures companies face in today's polarized environment and the challenges of navigating them. Stakeholders are pushing companies to wade into sensitive social and political issues, especially as they see governments failing to do so effectively. As CEOs, we do not always get it right. Moreover, what is appropriate for one company may not be for another.

One thing, however, is sure the world needs your leadership. As divisions continue to deepen, companies must demonstrate their commitment to the countries, regions, and communities where they operate, particularly on issues central to the world's future prosperity. Companies cannot solve every issue of public importance, but there are many – from retirement to infrastructure to preparing workers for the jobs of the future – that cannot be solved without corporate leadership.

Retirement, in particular, is an area where companies must reestablish their traditional leadership role. For much of the 20th Century, it was an element of the social compact in many countries that employers had a responsibility to help workers navigate retirement. In some countries, particularly the United States, the shift to defined contribution plans changed the structure of that responsibility, leaving too many workers unprepared. Moreover, nearly all countries are confronting greater longevity and how to pay for it. This lack of preparedness for retirement is fueling enormous anxiety and fear, undermining productivity in the workplace, and amplifying populism in the political sphere.

In response, companies must embrace a greater responsibility to help workers navigate retirement, lending their expertise and capacity for innovation to solve this immense global challenge. In doing so, companies will create not just a more stable and engaged workforce, but also a more economically secure population in the places where they operate.

A New Generation's Focus on Purpose

Companies that fulfill their purpose and responsibilities to stakeholders reap the rewards over the long-term. Companies that ignore them stumble and fail. This dynamic is becoming increasingly apparent as the public holds companies to more exacting standards. Furthermore, it will continue to accelerate as millennials – who today represent 35 percent of the workforce – express new expectations of the companies they work for, buy from, and invest in.

Attracting and retaining the best talent increasingly requires a clear expression of purpose. With unemployment improving across the globe, workers, not just shareholders, can and will have a more significant say in defining a company's purpose, priorities, and even the specifics of its business.

Over the past year, we have seen some of the world's most skilled employees stage walkouts and participate in contentious town halls, expressing their perspective on the importance of corporate purpose. This phenomenon will only grow as millennials, and even younger generations occupy increasingly senior positions in businesses. In a recent survey by Deloitte, millennial workers were asked what the primary purpose of businesses should be. Sixty-three percent of them said, "Improving society," while the rest said, "Generating profit."

In the years to come, the sentiments of these generations will drive not only their decisions as employees but also as investors, with the world undergoing the most massive transfer of wealth in history. $24 trillion from baby boomers to millennials as wealth shifts and investing preferences change, environmental, social, and governance issues will be increasingly material to corporate valuations. This is one of the reasons why BlackRock devotes considerable resources to improving the data and analytics for measuring these factors, integrates them across our entire investment platform, and engages with the companies in which we invest on behalf of our clients to better understand your approach to them.

BlackRock's Engagement in 2019

BlackRock's Investment Stewardship engagement priorities for 2019 are governance, including your company's approach to board diversity, corporate strategy, and capital allocation. The compensation that promotes long-term environmental risks and opportunities and human capital management. These priorities reflect our commitment to engaging around issues that influence a company's prospects not over the next quarter, but over the long horizons that our clients are planning for.

In these engagements, we do not focus on your day-to-day operations. Instead, we seek to understand your strategy for achieving long-term growth. Moreover, as I said last year, for engagements to be productive, they cannot occur only during proxy season when the discussion is about an up-or-down vote on proxy proposals. The best outcomes come from a robust, year-round dialogue.

We recognize that companies must often make difficult decisions in the service of broader strategic objectives. For example, whether to pursue certain business lines or markets as stakeholder expectations evolve, or, at times, whether the shape of the company's workforce needs to change. BlackRock itself, after several years of growing our workforce by 7 percent annually, recently made reductions in order to enable reinvestment in talent and growth over the long term. Clarity of purpose helps companies more effectively make these strategic pivots in the service of long-run goals.

Over the past year, our Investment Stewardship team has begun to speak to companies about corporate purpose and how it aligns with culture and corporate strategy, and we have been encouraged by the commitment of companies to engaging with us on this issue. We have no intention of telling companies what their purpose should be. That is the role of your management team and your board of directors. Instead, we seek to understand how a company's purpose informs its strategy and culture to underpin sustainable financial performance. Details on our approach to engaging on these issues can be found at BlackRock.com/purpose.

I remain optimistic about the world's future and the prospects for investors and companies taking a long-term approach. Our clients depend on that patient approach in order to achieve their most important financial goals. And in turn, the world depends on you to embrace and advocate for a long-term approach in business. At a time of great political and economic disruption, your leadership is indispensable.

Sincerely,

Larry Fink

Chairman and Chief Executive Officer

Laurence D. Fink is Founder, Chairman and Chief Executive Officer of BlackRock, Inc. He also leads the firm's Global Executive Committee.

9. GOVERNMENT'S DEFENSE – JOB DISRUPTING AI, ROBOTS, & MACHINES

The singular most significant change to the heralded rise of AI in China was the direction its government decided to take. China went from a position of nowhere near the AI A-list to their present position, where they are now competing with the United States for global AI dominance. The Chinese AI story is a testament to what is achievable if the government of a country decides to create an AI-first plan to drive their future. Going from obscurity to becoming a global force for AI is only as complicated as the ability to properly marshal resources and policies and a definitive long term strategy for the growth of AI.

For government, the path to a successful AI policy must be charted across terrains ranging from empowering its scientific communities and incentivizing research, providing enabling environment for talent development, building up capacity for skills and education, breaking down barriers for public and private sector adoption of AI products and services, balancing out issues relating to ethics and inclusion, providing equitable standards and regulations, and investing in data as a new resource while providing digital infrastructure for the entire system to remain sustainable and scalable over time.

Artificial intelligence should be seen as the one unique technology that should be synchronized with the country's economic plans and projections. The government must take up an active role both as an end-user and as an enabler of AI as a technology. Unlike the case for nuclear power where a nation may decide against its adoption due to ethical or safety reasons, AI does not give room for alternatives. Like electricity, it will stand as the new defining ingredient for advancing civilization to a whole new era. As a nation, you either develop your own unique AI strategy, or you lose global relevance at the table where the AI superpowers meet. In the race for global AI domination, early adopters will enjoy some unique dividends. However, the real trophies will be counted by the nations who can win emphatically in the data collection and data management race. For AI supremacy; Algorithms, AI infrastructure, and AI talents would be necessary, but none would be as important or hold the same weight, which the advantage of qualitative and quantitative data brings to a nation's AI game. So, while first movers will have some unique advantages, there would still be enough room for latecomer nations to catch up, especially if they do good work in optimizing their data gathering and management processes.

An even more significant advantage will be gained by populous nations with the natural disposition to churn out data in size. The Chinese AI story provides a perfect illustration of what the global AI race may look like. While China may have had a good foot-down on technology and industrial processes in the last decade, they were nowhere close to relevance on the global AI chart. Their rapid rise to AI dominance – where they now challenge the global leader, the United States – almost seems miraculous. It is almost inconceivable to realize that China released its AI policy plan in 2017 (less than three years ago from the time of writing this book in 2019). From that time till date, the transformation in their race for global AI domination appears to have been burning on nitro. There is much to learn from the Chinese AI advancement story. To date, their national AI policy remains the most comprehensive and focalized. The document, which was titled, 'A Next-Generation Artificial Intelligence Development Plan,' provides a sustainable framework that will help China in the areas of R&D, industrialization, talent development, education and skills acquisition, standard-setting, and regulations, ethical norms, and security. The document also sets some global 'table-shaking' benchmarks by specifying that China will become the "primary" center for AI innovation by 2030. One other thing that sets the Chinese AI policy plan apart from that of other nations is how it conjoins the development of its AI technology with specific economic milestones. According to the Chinese AI development plan, the government of China aims to cultivate an AI industry worth 1trillion dollars, with related industries worth 10trillion dollars by the year 2030.

According to Kai-Fu Lee, in his book AI Super-Powers, he asserts, "I believe that China will soon match or even overtake the United States in developing and deploying artificial intelligence."

While the race for global AI domination appears to be between the US and China, many other countries are also putting their best foot forward with their unique documentation of AI plans, policies, and strategies. Within only a two-year frame, the list of countries with documented AI strategies has now grown to 26 in number. It is truly a very interesting list, as you will find countries like Kenya and Tunisia listed alongside the United States, China, and even the EU Commission.

LIST OF COUNTRIES WITH DOCUMENTED AI POLICY AS OF 2019

1. Australia
2. Canada
3. China
4. Denmark
5. EU Commission
6. Finland
7. France
8. Germany
9. India
10. Italy
11. Japan
12. Kenya
13. Malaysia

14. Mexico
15. New Zealand
16. Nordic-Baltic Region
17. Poland
18. Russia
19. Singapore
20. South Korea
21. Sweden
22. Taiwan
23. Tunisia
24. UAE
25. United Kingdom
26. United States

While it is essential for every nation to have their name on this list, it must also clearly be understood that they can only win when their AI policy document and strategy are tailor-fitted to address their own unique situation. In the race to global AI supremacy, 'Charity must begin from home.' Every nation will have some nuanced approach to how they address their AI strategy. One factor that will determine this shade across different countries and cultures will be in the flavor of their data. Every country will have a unique data flavor that gives a different taste to the kinds of applications they build and how its citizens will end up interacting with AI solutions.

So the government will need to do their part in making it easy for AI researchers to acquire local data; it is this that will lead to the rapid iteration of solutions designed to solve local problems. AI solutions that are efficient in addressing farming problems in the US may find itself out of place when brought to African nations. Advancement in areas such as autonomous vehicles, robotics, and automated production processes should not be seen as a milestone that has been achieved on behalf of every country. African countries like Nigeria still lacks quality infrastructure like roads and electricity, so an autonomous electric vehicle will be entirely out of place until these problems are resolved. With a rising population number and an equally low unemployment rating, developing countries like Nigeria need to find creative ways in which their AI policy document can limit the loss of jobs due to AI and robot automation. AI opportunities for developed countries will be different for developing countries. Each country needs to find a specific model that addresses its unique situation.

Generally speaking, two salient policies that will determine how quickly nations climb the global AI chart will be for policies that address privacy, and policies that address how efficiently the government can make data available for researchers – private and public sector developers and entrepreneurs in general. This will be a very tricky water to navigate. Against the credit of China, it had to sacrifice hugely on the user data privacy side so it could gain on the ease of data availability side. On the other hand, a country like the United States and, more specifically the European Commission is making sure to prioritize security and privacy for their citizens' data over how easy it is to obtain data for AI product development. Each approach has its pros and cons, but whatever approach a country decides to prioritize, they must do so based on globally acceptable standards of ethics and the collective good of its citizens. For many nations, the primary driving force for putting together a comprehensive AI plan going into the future is due to the fear of job loss for its citizens and the 'Fear of Missing Out' (FOMO) in the global race for AI dominance. These are both very valid fears, and every government should take them seriously. Even amongst issues such as ethical concerns for countries who would try to use AI as a tool for warfare or humanity's eventual creation of a sentient AI that may destroy the human race, the issue of job loss due to AI and robot automation remains a recurrent discussion amongst global leaders and one of the chief reasons why many nations are rushing forward to put together their AI and digital policy plan for the coming years.

ADDRESSING THE CRISIS OF AI and JOB LOSS DUE TO AUTOMATION

With the advancement of AI in areas such as Deep Learning and Reinforcement Learning, the narrative of the threat that AI holds is quickly shifting from that of a sentient killer robot to one of robots with the capability to displace us from our place of work. This fear becomes even more lethal once we realize it comes as a two-edged sword. While billions of jobs are being wiped out, at one end, there will also be the problem of accelerated difference in inequality on the other end of the scale. Some economic and political experts are already forecasting dire glooms if the employment rate drops at hard to control rates as projected. First, most democratic systems of government will need to evolve into shades of authoritarianism in order to keep their restive population in check. Furthermore, as inequality becomes steeper in contrast, it would naturally result in elitism. The case for many communities will be that of the wealthy elites living in heavily secured and gated communities, while the 'unfortunates' squalor in their shantytowns and survive based on subsidies. This is already a picture painted in some African countries, e.g. Nigeria. So, the idea that AI and robots can effectively automate the painting of such kinds of gloomy pictures, even in already balanced democracies and systems around the world, has been a wakeup call to many nations.

To governments around the world, unmanageable unemployment levels and unfair inequality are the perfect ingredients for anarchy and a failed state. The job loss dilemma will not only cause negative economic disruption, but it will have the potential for quickly spiraling into something of an existential threat both domestically and between nations. There will be severe political repercussions and political alienations if the right proactive steps are not taken early on. The world before now may have enjoyed the benefits of globalization, but AI can quickly become an antithetical force to this ideal. Already, it seems like the bright summer of globalization is quickly splintering up into many small cold winters. Nationalism is on the rise, racial and religious discriminations are gaining new effrontery, and some of the biggest nations of the world are talking about building walls and or closing borders. The view from an economic perspective also has some fogginess in its lens; rise in trade wars and sanctions, increase in plagiarism and disregard for proprietary resources, and also the fear of technological domination which is stifling trade and openness between nations. So, for many world leaders, putting faith in a future world where the powers and opportunities of AI are openly shared is pure unfounded idealism. The reality of the situation is the very reason why many nations have put together their own distinct AI policy documents and plans. In fact, in a way, the AI policy document put forward by many countries can almost be seen as a manifesto for how they as a nation intend to survive the crash against the iceberg of AI's many jagged faces. Chief amongst these fears is the two-edged sword of unemployment and inequality.

While there is a need for a collective global strategy, many nations have realized that there is an even greater need for immediate action by their government to respond first at a national level. Each nation around the world needs to have a clear, well-laid-out strategy for dealing with the job loss problems caused by AI and robot automation. It falls squarely on the shoulders of governments to find the right balance for economic models that can create human-safe jobs considering the specific nuance of their economy. Government response to this situation cannot be a one-off kind of fix; it will require a cycle of planning, execution, evaluation, and reiteration. There will be the need for governments to create dedicated offices, institutions and parastatals that focus on encouraging R&D and overall advancement of the AI sector; while continuously iterating through solutions that address the negative fall-out of the technology. Every government will require long term plans for how they navigate the looming crisis of their citizens' displacement from work due to AI and robot automation. No government should be excused to let this issue manifest into an unplanned-for emergency in the future. They should have the foresight to anticipate the different shades of problems that may arise in the 2020s and the decades after them.

One sure strategy all governments should factor into their AI plans and policy is one that details how it would utilize education as a fortification against the army of AI and robot automation. Education and, more specifically, Re-Education will play a significant role in how nations can navigate the job loss and inequality threat at a national level. Dedicated educational and internship programs should be instituted, and orientation agencies should be mandated to carry out sensitization and awareness campaigns for all levels of society. Many schools still utilize curriculums and models that were designed for the industrial age – this needs to be overhauled quickly. The curriculum used in schools should be redesigned to accommodate digital skills as its core. Creativity, curiosity, care, empathy and all the other values that make humans distinct from hard, cold and calculating AI robots should be prioritized. No individual or level of society should be left out in this educational fortification process. It will be the only sure bet for nations who want to transcend in the future.

PAYING EVERYONE WITH ROBOT MONEY

The negative impact on the economy that would be caused by the AI and automated robot intrusion into the workplace is not one where the overall economic trajectory takes a deep dive: the direct opposite is the case. The overall trajectory of economic growth and GDP is forecasted to accelerate as AI, and automated robots help us do much more work in shorter amounts of time for a lesser cost. The only problem is that this accruement to the overall economic return will not be evenly distributed. Only those who have the capital to employ and maintain robot operations at scale will have the opportunity to eat the food that these AI and automated robots bring to the table. So, while the robots negatively impact the lower half of society by retrenching workers and contracting their economic power, it enthrones another class by giving them a perpetual cycle of making wealth. As I have discussed before, no democratic government or even any other form of government would conscientiously allow its society to fall into such a state, not even a government based reasonably on fascism.

So the need arises for governments to step in and create systems designed to balance out this situation. Many proposals have been made for how this problem should be addressed. Thought-leading experts like

Bill Gates have lent their voice to the discussion by proposing models where robot and robot employers are taxed, and the funds redistributed to those on the unfortunate side of the continuum caused by robot displacement.

One other proposal that is becoming popular amongst industry experts propounds the idea that AI and robot automation themselves should be seen as resources that should be evenly distributed. So if the rich have access to AI tools and robot processes that help them create more wealth, those of lesser means should also be given an equal opportunity to explore their ideas by utilizing the power of AI tools and robot processes. The argument for this approach is based on the idea that AI should be seen as providing the same opportunities as electricity, so it must be made available to everyone. No doubt, for such an idealistic solution to be possible, external regulators in the form of government agencies will need to work out all the kinks to ensure fairness without appearing as though the government itself is spiraling into a form of communism.

UNIVERSAL BASIC INCOME (UBI)

Of all the proposals put forward to address the job crisis of the future, the Universal Basic Income (UBI) approach has, by far, received the highest amount of applause. It is already being test run at different scales and models by some countries around the world. Even more practically, a candidate running for the 2020 President of the United States, Andrew Yang, has made it central to his campaign manifesto.

SO WHAT EXACTLY IS THE UNIVERSAL BASIC INCOME, AND WHY IS IT POPULAR?

The idea of UBI at its core proposes the provision of regular income stipends to citizens of a country or members of a community by a government or social organization. The popular models of UBI advocate that everyone, regardless of whether they are employed or not, regardless of their economic status or social class, should benefit from the monthly or yearly payment of the UBI scheme. Whether this is a good idea or not, has been the subject of numerous research and experimentation by economists, non-profit organizations, and countries around the world. The idea of UBI is not new to this day and age; it can be traced even to the time of the Roman Empire, where food guarantees and stipends were given to people in order to maintain the economic equilibrium in its cities.

In recent centuries, the UBI ideal has continued to receive approval and push from a wide variety of people, such as revolutionaries, politicians, economists, entrepreneurs, and social reformers. As far back as the 18th century, people like the English-American revolutionary, Thomas Paine, believed everyone above the age of twenty-one should receive a payment as compensation for the loss of their natural inheritance. Also, Nobel Prize-winning economists like Milton Friedman and F.A. Hayek believed the idea of a periodic minimum income would be a force of greater good for society in general. According to F.A. Hayek an Austrian economist,

"The assurance of a certain minimum income for everyone, or a sort of floor below which nobody need fall even when he is unable to provide for himself, appears not only to be a wholly legitimate protection against a risk common to all but a necessary part of the Great Society in which the individual no longer has specific claims on the members of the particular small group into which he was born."

Even more convincing of the trust put on the UBI model, was the fact that during the 1960s, Martin Luther King Jr. joined over 1,000 economists from different universities across the United States to sign a letter advocating for this model. This letter was then presented to President Nixon in advocacy for an income guarantee irrespective of race or economic class. Under the Nixon administration of the 1970s, this idea of providing everyone with a guaranteed income was almost passed into law.

Today, the debate behind whether everyone should receive payment in order to cover their basic cost of living and provide them with financial security; has regained heightened popularity as the army of AI and robot automation prepare to disrupt workplaces and workers' income. Today, the leading voices in the conversation of UBI have moved from social-justice fighters, economists, and politicians to tech entrepreneurs, founders, and Silicon Valley businessmen – the very people empowering the AI and robot army to cause onslaught in workplaces and society at large. Facebook's founder, Mark Zuckerberg, amongst other leading tech voices like Elon Musk, Peter Thiel, Ray Kurzweil, and even Virgin Atlantic CEO, Richard Branson, all believe the only way the ship of society can navigate the murky waters of the future is through the implementation of a form of the Universal Basic Income. In a global economic system that is increasingly becoming more capitalistic, is it safe to assume this advocacy for UBI by tech leaders is genuinely altruistic?

According to Kai-Fu-Lee, in his book, AI Super-Powers, he believes he understands why the Silicon Valley elite have taken favor to the idea of UBI. ***"It is a simple, technical solution to an enormous and complex social problem of their own making."***

He goes further to say, ***"In observing Silicon Valley's surge in interest around UBI, I believe some of that advocacy has emerged from a place of true and genuine concern for those who will be displaced by new technologies. However, I worry that there is also a more self-interested component: Silicon Valley entrepreneurs know that their billions in riches and their role in instigating these disruptions make them an obvious target of mob anger if things ever spin out of control. With that fear fresh in their minds, I wonder if this group has begun casting about for a quick fix to problems ahead."***

I believe that Kai-Fu Lee's assertion has some validity. Nevertheless, at this point in time, more focus should be geared towards harmonization for a comprehensive solution to the looming job crisis and inequality problem as the army of AI, and automated robots keep on marching towards us.

MAKING UBI POSSIBLE

The Universal Basic Income (UBI) program, no matter how forceful it is being heralded as the ultimate solution to the problems of the future, cannot be successfully implemented independently of government involvement. The right approach should see government at the forefront of the battlefield, supported by financial artilleries from private or public businesses, while receiving aerial support from independent organizations like the United Nations and other third party institutions invested in socio-economic development. Every government needs to put itself in a place of total responsibility for how effectively it has prepared against this crisis. Technology and businesses will continue to disrupt society as there is little or nothing governments can do in limiting the trajectory of technology's exponential climb. Any policing and limitation in front of technological advancement will result in a direct inhibition to civilization and the stifling of humanity's progress for reaching its true potential. The dark ages provide experiential proof of this.

If economists, futurists, entrepreneurs, people in business, and politicians have all given credence to UBI as the most effective ground to stage the war against the army of AI and automated robots, then governments must waste no time in beginning the build of entrenchments and forts on this battleground. The perfect UBI weapon must be built early-on in this campaign, and it must be built before we start seeing the full effects of massive job displacement, uncontrollable unemployment and unmanageable inequality. While independent institutions like the United Nations, through its sustainable development programs, are directly and indirectly trying to add some firepower in the fight for equality and financial inclusion, some governments around the world have begun test-running different UBI models that can work for its citizens. Countries like Canada, India, France, Finland, Kenya, Switzerland, Scotland, and more have either considered implementing it or are currently running pilot programs on it.

TESTING OUT THE FIREPOWER OF UBI

- *In 2017, Finland began a two-year experiment. It gave 2,000 unemployed people 560 euros a month for two years.*
- *Holland is running a pilot program in Utrecht, where it pays 250 people 960 euros a month.*
- *Canada is experimenting with a basic income program. It will give 4,000 Ontario residents living in poverty C$17,000 a year.*
- *In 2017, Kenya announced a 12 year pilot to benefit 6,000 villagers. They will receive a $22 monthly payment on their smart phone equivalent.*
- *The government of Scotland is currently running a research program that will inform them on how best to provide a minimum that pays every citizen for life. The program projects paying retirees 150 pounds a week. Working adults would be paid 100 pounds and children below 16 would be paid 50 pounds every week.*
- *Taiwan is considering voting on a basic income for its citizens. It is believed this will help them stop the massive emigration of its workforce and help its senior citizens, who are mostly left behind to live in poverty.*
- *Stockton, California, is planning a two-year pilot program for fall 2018. It would give $500 a month to 100 local families. It hopes to keep families together, and away from payday lenders, pawn shops, and gangs.*
- *Chicago, Illinois, is considering a pilot to give 1,000 families $500 a month.*

As each country or city seeks to find a UBI model that best works for it, there will be a need for its government to adopt the Silicon Valley approach to building and scaling projects. Governments will need to start learning lean systems of operation and agile methods of project development. They will need to learn that a defining skill of success in the future is the skill of failing fast and learning as you go. A UBI model that works perfectly for one country may be a disaster for another. So, every government needs to start researching and prototyping what works best for it. They will need to learn how to do this fast and without the impediment of unnecessary political bottle-necks or snail-paced administrative bureaucracies.

THE CASE AGAINST UTOPIA

Many of the economic experts, who argue against UBI, believe it sounds great only because it is utopian. Some believe it could end up becoming a Trojan horse that stifles natural economic growth. One of the biggest arguments against those evangelizing for UBI is that it does not come without some unique set of challenges. Chief amongst this is the fear that; giving everyone money for doing no work will uncontrollably increase their spending power. Furthermore, if things get to a point where people's demand for products and services can no longer be met by supply from manufacturers and service providers, then, inflation becomes the result. Inflation, on the other hand, naturally leads to higher prices of goods and services, which can only be afforded by those with more money. Over time, this can create an even broader economic divide, which becomes fuel to the growing fire of inequality.

There is also the fear that many of the citizens will see no reason to invest in building their skills over the long run, and therefore remain perpetually unemployable. While those who may have some employment will place low value on hard work as long as they are sure that they would be getting a recurrent paycheck every month. As fewer people are willing to participate actively in the labor market, there would be an even higher need for more automation and more dependence on the robots to do more of the jobs that humans have on their strongholds. This is a worrisome cycle many experts fear UBI might suck us into.

Lastly, where will the money come from? Many of the UBI evangelicals propose that every robot or automated process should be taxed, Value Added Tax (VAT) should be increased, and funding for other social projects converted directly to cash. All of these proposals do not provide a comprehensive model for funding the UBI program. Moreover, even if these suggestions get to work out in some countries, they may not be feasible in others.

THE CASE FOR UTOPIA

As loud as the critics of UBI are making their voice sound, it is continually drowned by the voices of those who believe UBI is not only a means to address the issues of job displacement and inequality, but that UBI may also end up becoming the silver-bullet that solves many of society's malady.

According to research, 36% who currently have a job are underemployed, and around 38% are not happy with the current job they are doing.

With UBI, many of these people will have a financial cushion and time to upskill or reskill themselves for the jobs they would sincerely be passionate about doing. UBI evangelists believe that with the financial cushion that UBI will provide, more people will explore entrepreneurship or become engaged in volunteer work that benefits society at large. UBI will also offer some unique advantages to the government. It will help them minimize bureaucracy and reduce administrative costs when compared to its other myriad traditional welfare programs. Overall, apart from the bottleneck of funding it, the UBI remains as an appealing program for governments around the world. It is like that one medicinal tablet that can help it cure almost all of its maladies – all of these without any change in government itself or the system it operates. UBI is also a desirable solution to the employees of labor. It absolves them from having to invest too much in workers' welfare or worrying too much about the impact of their profit-oriented automation processes. For product manufacturers and service providers, there is also the guarantee of demand for their supply.

One of the biggest arguments put forward for UBI as the ultimate solution to the job crisis and inequality problems is that it would provide the opportunity for humans to be more human. When we no longer have to worry about work and income, we will direct our minds to family and community building. We will have the time and opportunity to explore the things that make us truly happy – creativity, arts, travel, and overall well-being of humans will help us transcend into a new era. Many of the supporters of the UBI believe it is the one solution that can help humans achieve the closest ideal of utopia we can get to. It will help us flower a new kind of culture and create more robustness and expansion of the economy. There will be an increase in art, music, and innovations. It will help us change the narrative behind politics and government, racism and inequality will be given little or no chance to fester, humans will have the opportunity to redefine the meaning of work, life and community.

FINAL THOUGHTS .

A tremendous determining factor for how we can chart the course of human progress through this storm of AI and robot automation will depend on the roles, policies, and effectiveness of government as they prepare to meet this situation. Any government that wishes to be successful in the era of AI and robot domination must know how to differentiate the rose flower from its thorny stems. AI should not be regarded as a threat to be mitigated, instead, it should be seen as an opportunity to be capitalized on. With the right strategies and policies in place, governments around the world can use the power of AI and automation to position themselves for prosperity in the new age.

For example, "Accenture estimates that AI could double economic growth rates by 2035 and boost labor productivity by up to 40%".

The place for governments to start is at the policy level: favorable policies like the creation of import barriers and the provision of incentives for local manufacturing companies. This will be a crucial strategy, especially for developing nations and countries that want to compete globally by providing AI and automation as a service. The institution of import barriers will limit the importation of cheap products whose prices have been driven down due to automated production processes. This will then give local manufacturers some breathing space from being squeezed out of business. Also, policies should be put in place to provide subsidy for these local manufacturers, so they too can ramp-up their production processes by adopting AI and automated technologies. Surviving this wave of AI dominance will require tact, agility, and the willingness to keep iterating over policies and strategies until a solution that works for that country is gotten. One limiting factor I fear will slow progress for many governments will be the trap of exchanging empty political rhetoric between parties and groups in governments, and the unnecessary politicking of situations that demand quick rationale solutions. It may be quite unfortunate to live in a future where AI and robot automation lead to joblessness, inequality, and broken systems simply because the politics of men did not allow them to see beyond their ego or objectives of how they could get elected for another term in office.

If the government of a nation must win this war against the forward marching army of AI and automated robots, then it must reevaluate its primary function as the machinery needed to drive society towards a better ideal. The government will need to see itself in a new light – in the light of being subservient to the collective will of the people it is sworn to serve. The only way the government can survive the storms that lay ahead in an AI and robot-automated future is for it to be true to its purpose of being in existence in the first place. A purpose which it must understands to mean - the unreserved and straightforward service for its citizens.

PART THREE

THE FUTURE OF HUMANITY

10.
GENESIS AND EXODUS - THE FUTURE OF HUMANITY

This chapter and the ones that come after it are going to be the most challenging sections to put together in the course of this book. The intention is to harmonize the many thoughts across science, philosophy, and religion, and see if they can provide a foundation for understanding what was, what is, and what will be. Before we begin our journey through the maze of reason posed in this section of the book, I think we must agree on the fact that it will be impossible to safely chart a course ahead for humans without first finding an adequate foothold in the century-old debate of man's origin.

Now, it is not my intention to take any side of the debate on whether humans were created by an intelligent being 'a creator- God' or if man is simply the result of the Darwinian evolutionary process that bases its proof on carbon prints spanning millions of years. Since Charles Darwin published his work, The Origin of Species in 1859, the one-sided religious narrative generally used for describing humans – their origin, purpose for being, and especially the future for the human race – has branched out into many spectrums of interpretations, some of them compellingly unreal, some of them conceivable only through complex mathematics, while some of them logical by every means but sounding stranger than fiction. There are a thousand and one books, debates, and publications where experts from all sides of this rift have put their best foot forward;

so our assumptions as we explore this maze in search of the outlets that lead to the future of the human race will be based on narratives guided by research and the experienced views from sages in the field of science, technology, philosophy, and religion.

We will have to go through the non-refutable arguments of the Oxford University professor and philosopher, Nick Bostrom, whose 2003 paper 'Are We Living In a Computer Simulation?' caused quite a stir in the academic world, and the ripples of that research work has filtered into many of the transhumanist discussions and projections for who or what humans are, and what they should or will become. Advancement in artificial intelligence also adds its unique flavors and coloring to the already spicy debate of where we need to peg our origin and where we need to set our compass for in the future. Will Artificial Intelligence grow into Artificial General Intelligence (AGI) and eventually lead us to the Singularity? According to some of the significant evangelists for the Singularity such as; Ray Kurzweil and Peter Diamandis, the question is not 'if' but 'when.' These and many other thought-leading experts who have spent decades developing and advancing technologies in the field of AI believe we are already about to cross the Rubicon because the Singularity is already on its way by just a few decades ahead of where we are at the moment. The idea that machine intelligence is already on its way to surpassing human intelligence and that, humans, may eventually end up taking their creation of AI as some god is one that many futurist have had to factor-in as they try to piece together the bits and pieces of the vehicle that will drive humans to their future.

Technology entrepreneurs like Elon Musk have even taken the discussion farther by implying that we will have no option but to merge and evolve with these machines. His investment of $100 million dollars in the company NeuraLink, where he is majority shareholder, is proof that this is no mere assertion or publicity scheme.

Nevertheless, all of these futuristic ideals by tech entrepreneurs and researchers can easily be complicated by the over five billion people globally who adhere to one religious tenet or the other. For example, a majority of the over two billion Christians worldwide believe not only in a supremely intelligent creator - God. They also believe we currently exist in some form of simulation where those who meet the set requirement will be saved from the system of this world and set for life in eternity. They look forward to the time of Armageddon (end of the world as espoused in the Bible's book of Revelation), which is believed will occur in the not too distant future.

Over one billion Muslims worldwide also have their own general narrative for what the hereafter will be. In the Muslim text, "Jannah," – which is the paradise that sums up the hope of every Muslim believer – is described with real delights, such as beautiful maidens for men and young men for women, precious stones, delicious foods, and continuously flowing water. The Islamic text describes life for its immortal inhabitants as one that is happy—without hurt, sorrow, fear, or shame—where every wish is fulfilled. While this view of eternity for the faithful is as utopian as it gets, the eschatology painted for non-believers and the world in judgment is one as graphic as the Christian Armageddon if its religious text is to be taken literally. The Jews, Hindus, Buddhists, and many of the other eastern religions also have varying futuristic ideals for what man and the state of the world will become in the future.

The big caveat, to all these beliefs, is that none of them aligns in any way with the pictures painted so far by technologists and futurists such as Ray Kurzweil or Elon Musk. Even for the technologists who do not believe in ideals such as the singularity or machine intelligence overtaking human intelligence, there still exists a significant diversion in thought between them and any of these religious views of what the future of man and the world will be. We remain in this impasse if we keep trying to juxtapose any of these religious views with those held by the scientific and technological community. So the question becomes, why is it even necessary that we find a harmonization or middle ground that can bridge these two extreme dialectical positions sustained by religion on one end and heralded by advancing science and technology on the other end?

To answer this question, we find that in a world of over seven billion occupants, the overshadowing majority (over 80%) subscribe to one form of religious belief or the other. Moreover, we can almost say for sure that the vast majority of the people whose life view is shaped by religious sentiments approach many of life's questions solely by the response of faith, barring reason, and scientific revelations in many occasions. As science and technology advance us into a new era, the most significant limiting factor may not be from humans' ability to keep developing these technologies. It may come from the superstitions and prejudices from billions of religionists whose perception of the world is shaped mainly by the handed-down traditions and oracles from long-forgotten eras. It will require concentrated effort, patience, and creativity to usher a future of almost limitless possibilities to people whose present reality is constricted by many unfounded frameworks that have shaped their lives and thinking from birth.

Almost all of the world's religion is designed with a top-down structure where traditions, doctrines, and dogmas are passed down to members without the members having any say to oppugn the how and why of such tenets. All that the Pope, Pastor, Imam, Rabbi, Prophets, Monks or whatever figure these religious adherents often look up to for the direction of their faith needs to do, is to place technological advancement in a disparaging position by preaching, teaching or throwing anathemas against it and the result will be millions of people who are ready to burn, exterminate and take up a war against the heathen curse of technology with its creator. A situation where health and life are jeopardized due to religious sentiments against abortion, blood transfusion, or vaccination, is not uncommon. The world with all its knowledge and progress over the past few centuries still have situations like the stamping of sin on formal education or placing an embargo on girl child education in some places where the religious adherents are only concerned with preserving the traditions of their fathers and not sinning against their deity. Whether their actions cause the stalling of scientific progress or cause an uneven distribution of technology is of little consequence to them, they only see as far as their religious or communal head can.

WHERE IT ALL BEGINS

How did the view we have of the world get so fragmented and so divergent in positions that it now becomes a potential stumbling block for the progress of civilization and the future of the human race? Well, the good news is that both science and religion all agree that beyond every reasonable doubt, our universe has a beginning. Furthermore, that beginning started by a force alien to all our present understanding.

From this early congruence in construct for a universe that started from something alien, is it possible for both science and religion to also trace and harmonize an understanding for the future of the universe? In the following sections of this chapter, we will explore some ideas and thoughts on this, but for now, let us see where science and religion diverge.

FROM THE VIEW OF SCIENCE

For science, the narration began around 13.8 billion years ago, and this narration starts conventionally with a Big Bang! The big bang theory postulates that the reality of our universe started from the point of dense and hot super force, also known as the 'Singularity.'

So what was this point of singularity? Where did it come from? What is its nature? Unfortunately, these are all questions beyond the realm of science as we know it today. According to many of the science experts who have explored this topic, Singularities are zones that defy our current understanding of physics.

The Singularity In Cosmology – *Singularities are thought to also exist at the core of "black holes." Black holes are areas of intense gravitational pressure. The pressure is thought to be so intense that finite matter is squished into infinite density (a mathematical concept which truly boggles the mind). These zones of infinite density are called "singularities." Our universe is thought to have begun as an infinitesimally small, infinitely hot, infinitely dense, something - a singularity. Where did it come from? We do not know. Why did it appear? We do not know.*

Back in the late '60s and early '70s, when men first walked upon the moon, three British astrophysicists, Steven Hawking, George Ellis, and Roger Penrose, turned their attention to the Theory of Relativity and its implications regarding our notions of time. In 1968 and 1970, they published papers in which they extended Einstein's Theory of General Relativity to include measurements of time and space. According to their calculations, time and space had a finite beginning that corresponded to the origin of matter and energy. The singularity did not appear in space; instead, space began inside of the singularity. Before the singularity, nothing existed, not space, time, matter, or energy - nothing. So where and in what did the singularity appear if not in space? We do not know. We do not know where it came from, why it is here, or even where it is. All we know is that we are inside of it, and at one time, it did not exist, and neither did we.

So taking these generally accepted views from science about the start of our universe, the best conclusion we can come to is that an alien force birthed our universe. Oddly enough, when this understanding is juxtaposed alongside tangled concepts like quantum mechanics, string theory or even Nick Bostrom's theory of a Simulated Reality – as scientific or philosophical models to explain our present reality – science as it is today only generates even more deeper questions whose answers may begin to lead the searcher to spiral into the realm of metaphysics. Unfortunately for the pure believer of science, as soon as the word metaphysical or supernatural is invoked into any mysterious phenomenon that beats the rationality of science, the religionists and their idea of creationism like to quickly jump in and lead the direction of the debate.

FROM THE VIEW OF RELIGION

The generally accepted view of the religionist of how we got here is one based on faith, as revealed through divinely inspired teachings and writings. An infinitely powerful, infinitely conscious, and infinitely intelligent being created the reality we call our universe. For the religionist, there is no reason for conjecture – for time, space, and matter, are all part of an intelligent design intended to sustain the reality of our existence. The religionist point of view also gives a rationalization for our existence as a purpose-driven one. Humans were designed in the image and characteristic of this intelligent being, with the power to think and to do, to create and to spread positive energy 'entropically.' The religionist view, primarily held by Christians, provides a rationale for the present state of the world (with all its seeming chaos) and also sums its narrative as the result of the great controversy between two alien intelligence; two opposite ideals in a struggle for the fate of the universe.

The writer, Ellen G.White, sums up the belief shared by many of the faithful across multiple religions with these closing paragraphs to her iconic book, The Great Controversy.

"There [sometime in the future of the human race and the universe], immortal minds will contemplate with never-failing delight the wonders of creative power, the mysteries of redeeming love. There will be no cruel, deceiving foe to tempt to forgetfulness of God. Every faculty will be developed, every capacity increased. The acquirement of knowledge will not weary the mind or exhaust the energies. There the grandest enterprises may be carried forward, the loftiest aspirations reached, the highest ambitions realized; and still there will arise new heights to surmount, new wonders to admire, new truths to comprehend, fresh objects to call forth the powers of mind and soul and body.

All the treasures of the universe will be open to the study of God's redeemed. Unfettered by mortality, they wing their tireless flight to worlds afar--worlds that thrilled with sorrow at the spectacle of human woe and rang with songs of gladness at the tidings of a ransomed soul. With unutterable delight the children of earth enter into the joy and the wisdom of unfallen beings. They share the treasures of knowledge and understanding gained through ages upon ages in contemplation of God's handiwork. With undimmed vision they gaze upon the glory of creation--suns and stars and systems, all in their appointed order circling the throne of Deity. Upon all things, from the least to the greatest, the Creator's name is written, and in all are the riches of His power displayed.

And the years of eternity, as they roll, will bring richer and still more glorious revelations of God and of Christ. As knowledge is progressive, so will love, reverence, and happiness increase. The more men learn of God, the greater will be their admiration of His character. As Jesus opens before them the riches of redemption and the amazing achievements in the great controversy with Satan, the hearts of the ransomed thrill with more fervent devotion, and with more rapturous joy they sweep the harps of gold; and ten thousand times ten thousand and thousands of thousands of voices unite to swell the mighty chorus of praise.

"And every creature which is in heaven, and on the earth, and under the earth, and such as are in the sea, and all that are in them, heard I saying, Blessing, and honor, and glory, and power, be unto Him that sitteth upon the throne, and unto the Lamb for ever and ever." Revelation 5:13.

The great controversy is ended. Sin and sinners are no more. The entire universe is clean. One pulse of harmony and gladness beats through the vast creation. From Him who created all, flow life and light and gladness, throughout the realms of illimitable space. From the minutest atom to the greatest world, all things, animate and inanimate, in their unshadowed beauty and perfect joy, declare that God is love." – ***The Great Controversy – Page 678***

WHERE WE ARE NOW

Science and religion both have their places in society; they have become tools for humans as they try to unravel the secrets and meaning of their existence. However, one limiting factor that has plagued religion is that its rationale is often buried deep in too many superstitions and mysticism. Narratives based on simplistic faith do have their place in helping man deal with the reality of his existence. However, when this faith is placed in a non-progressive position, one that fails to rationalize revelations as revealed by science and technology, then such a faith becomes one whose house has been built on sandy foundation – for when the tsunami of progression in the form of science and technology washes upon it, it falls and often breaks into bits and pieces of bigoted fanaticism.

Over the centuries, religion has courted science, and the relationship has not always been a romantic one. For example, during the Middle Ages, when the Roman Catholic Church held almost complete control over the progress of science, technology, and astronomy, people like

Galileo Galilee were persecuted for proposing scientifically researched ideas on models for the world. The church taught and believed that every part of our solar system revolved around the earth.

Galileo Galilee, a renowned astronomer at that time, discovered this to be an erroneous ideology; with the aid of proficient telescopes and research, he discovered that the earth was not fundamental to the universe. Many of the scientists and astronomers of Galileo Galilee's time refuted this theory and plotted to have him punished for promoting such a blasphemous theory. He was later sentenced to life imprisonment. Below is an excerpt from a letter to his friend Kepler.

"My dear Kepler, I wish that we might laugh at the remarkable stupidity of the common herd. What do you have to say about the principal philosophers of this academy who are filled with the stubbornness of an asp and do not want to look at either the planets, moon, or the telescope, even though I have freely and deliberately offered them the opportunity a thousand times? Truly, just as the asp stops its ears, so do these philosophers shut their eyes to the light of truth."

Today we can easily criticize the folly of that majority of scholars, clergymen and prominent scientists who refuted the theories of Galileo Galilee, who rather than lend themselves to at least understanding what he was trying to say, closed all their faculties of reasoning because of traditions and belief in a fallacious doctrine. Today the general narrative has changed considerably from a pure religionist view to one where we can even say science and technology provide the general rationale that dictates the main ideologies being held by society. However, even as science and technology continue its march against the many superstitions and unfounded faith that seems to bedevil many of the religious ideologies around the world, it may have reached a point where a truce needs to be called. With revelations such as Quantum Mechanics, String Theory, and propositions such as Nick Bostrom's Simulation Hypothesis, science may have overreached itself, stretching its hands beyond the veil that seems to separate the physical from the metaphysical.

When philosophical debates that involve consciousness and the possibility of life after Artificial General Intelligence crosses the 'rubiconic-line' of the Singularity are added into this whole mix of man's origin, the rationalization of his present reality and the transcendence that will be his future, we are left with no option but to find that confluence between logic and instinct, objectivity and intuitiveness, knowledge and faith, science and religion. Can such a confluence exist? Can religion and science fit in the same tool shelf for helping us understand the many mysteries of our present reality, and safely help us in plotting or following after an already plotted pathway for our future? I believe these questions are essential. Even more so, I believe an answer that balances well on the fulcrum of such a confluence will provide us with the only rationale for understanding the great questions of who we truly are, the 'consciousness' of why we can ask why we exist, and our subconscious instinct to keep questioning what lies ahead for us in the future.

In the next section of this chapter, I intend for us to peer into what is, and what will be. We will utilize a philosophical binocular while standing on the summit of the confluence between science, technological progression, and religion. Religion doused with nonpartisanship, a religion whose faith can be substantiated by the revelations of science.

PROOF OF CONCEPT - AI PATHWAY TO GODHOOD

Sam Harris, an author, leading neuroscientist and philosopher, at one of his Ted talk made this profound statement, ***"The moment we admit that information processing is the source of intelligence, that some appropriate computational system is what the basis of intelligence is, if we admit that we will improve these systems continuously and we admit that the horizon of cognition far exceeds what we currently know, then we have to admit that we are in the process of building a god. Now is a good time to make sure it is a god we can live with."***

All of the hype and excitement surrounding artificial intelligence and all of the news-breaking headlines AI has caused so far are based on Narrow AI. Self-driving cars, autonomous drones, cancer diagnosis, etc.

are all powered by Weak AI or Narrow AI or Artificial Narrow Intelligence (ANI) – Which is an AI agent capable of performing and excelling only at a single task. The real debate begins when we enter the threshold of being able to make Strong AI or Broad AI or Artificial General Intelligence (AGI) – This is when AI becomes as intelligent as humans, having the capability to perform thinking and action as a human can. At this level, there is not much of a threat to human supremacy. However, according to projections, once AI attains the level where it is on the same intelligence level with humans, it becomes only a matter of a few years before it exponentially surpasses the intelligence level of humans. This is where the controversial region beyond which people like Sam Harris, as quoted above, warns. When AI attains the ability to function beyond the limit of human-level intelligence, it would have attained the status of Artificial Super Intelligence (ASI). While this is only theoretical at the moment, it is not impossible. There has been much debate on the timing or period when ASI may become a reality, some experts are very optimistic, and some are very pessimistic, but overall the vast majority of them agree that the path of technological progress through ASI is the most plausible future that can be deduced if our technology keeps improving as it is doing today.

Some technology and futurist thought-leaders are very confident in this coming future as to have begun investing ahead of time. We have people like Elon Musk sinking money into startups like NeuraLink (which hopes to connect the human brain in a non-invasive way to a computer), and on the other hand, you have people like Ray Kurzweil and Peter Diamandis who have opened a specialized training institute, The Singularity University (with the focus of preparing individuals and businesses for life and disruptions of the future).

The optimism and certainty expressed by these futurists, seasoned entrepreneurs, and technology experts are understandable given the fact that various technologies are beginning to collide with one another, and the results are not just bits and pieces of innovation alone, but sometimes one single collision can result in quantum leap for the overall progress of all technology. While some people would argue that Moore's law will reach its physical limit soon (when transistors can be made to operate at almost atomic levels), a lot of other people believe it has already given birth to enough descendants like the ability to fit a computer in your pocket many times the computing power NASA used in going to the moon, and the ability for the Internet to function as a global brain; one that is accessible virtually – connecting resources, ideas, communities and cultures that will continue to add up in the acceleration of growth and progress. What then happens when newer technologies begin to combine and copulate in ingenious ways with each other? What happens when quantum computing marries AI and flirts with nanotechnology while crushing on blockchain, robotics, and IoT all at the same time?

We have entered an era where we are sure only of uncertainties. Just like the disruption of blockchain and cryptocurrencies took the world unawares, no one can be certain what new technological innovation will help us fast-track or even quantum-leap technological progress into the age of Artificial General Intelligence (AGI) and the Singularity.

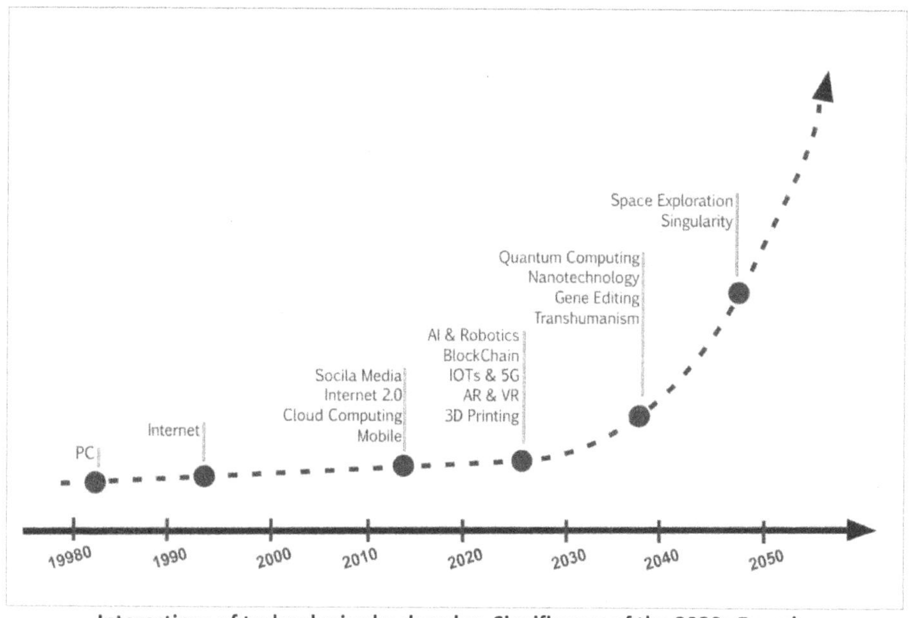

Interactions of technologies by decades- Significance of the 2020s Decade

The above diagram shows how crucial the era of the 2020s will be in defining the possibilities of the future. As various technologies continue to collide and interact, the only certainty futurist can guarantee will be that of uncertainties.

One of the leading columnists for Forbes, Naveen Joshi, shares in this optimistic-uncertainty that seems to cloud the coming decades, here is an excerpt from one of his articles:

"However, time-wise, the rapid rate at which AI is developing new capabilities means that we might be close to the inflection point when the AI research community surprises us with the development of artificial general intelligence. Moreover, experts have predicted the development of artificial intelligence [AGI] to be achieved as early as 2030, and the Singularity by the year 2060."

While in 'some' ways, I share in this optimistic realization of AGI by Naveen Joshi, I also realize that it is essential to understand the unique challenges that are strewn along the path to get there. Before anything else, we must be able to meet up with the computing powers required for AGI to run. If we use our human brain as a yardstick, then we are looking at computers that can carry out about ten quadrillion (that is a 10 with 16 zeros) computation per second. The human brain does have some considerable computing power, but as Moore's law and other technological breakthroughs emerge in the 2020s and beyond, we will be getting exponentially closer to arriving at the feat of creating the first general-purpose AGI level computer. The good news is that we currently have computers that have surpassed ten quadrillion calculations per second (10*16cps), and can even function at 34 quadrillion cps. The bad news is that there is only one such computer currently at the time of writing this book in 2019. Tianhe-2, a supercomputer owned by China, leads the way in terms of raw computing power. This one single computer cost $390 million to build, it uses 24 megawatts of power, and it is so big that it takes a space of around 720 square meters (almost three times the size of a tennis court). Considering the power usage of about 24MW in comparison to our brain's power usage of only about 20 watts implies there is still a long way to go in developing supercomputers capable of marching up with human intelligence. Nevertheless, just like the gigantic computers we had in the 1960s, whose size we can now fit into our phones, maybe the 2020s and decades beyond will give us computers with powers like the Tianhe-2 in the palm of our hands.

The path to AGI will not be an easy one, apart from the hardware and computing efficiency which computers like the Tianhe-2 are looking to pioneer, there remain issues like the need for advanced Software, Algorithms, Neural Networks and even areas like regulatory and ethical frameworks all need to take exponential leaps for us to arrive at the threshold of Artificial General Intelligence. Achieving AGI will be a real defining point for humans because the moment we step into the realms of AGI (which experts world-over have pegged tentatively will occur in 2045, roughly 25 years from now) we will have crossed a rubicon line from which the only way forward will be for us to march towards Artificial Super Intelligence (ASI).

Whenever the discussion moves beyond AGI to ASI, even the best of futurist begin to thread with caution. First, when we take into cognizance the fact that the only reason why humans as a species dominate the world and the natural order of things as it is today is due to their superior level of intelligence: then it becomes possible to state that if an intelligence –one that we created – surpasses human-level intelligence, this intelligence will have the upper advantage in usurping the mantle of superiority for the planet. There will be nothing to say that such intelligence cannot dominate the natural order of things just like humans can do presently. If dominance were based on size, elephants would have been the significant contenders; if based on speed, then the cheetah would have held sway over the planet, and the lion if dominance was to be based on sheer strength. Nonetheless all of these seeming physical limitations, man has continually sat at the head of the food chain, simply because of its higher intelligence. Man can build submarines and explore the depths of the sea; man can build skyscrapers taller than the tallest trees; man can build jets that can climb to altitudes surpassing even the highest mountain, if humans with their present level of intelligence can achieve all of these, then what limits us in saying that, an intelligence greater than ours will not take the same path as we have – the path of bending the natural order of things to meet its end. A careful look at the intelligence chart shows humans are only a fraction of steps ahead of apes one of the only animals that come close to it in terms of intelligence level, yet the impact of that small gap has ensured such a wide rift between the human world and the world of apes. The ape for all its intelligence and for all the training we can give to it will never understand the how and the why of a phone or a plane. It cannot even comprehend the complexity of our language or the soulful, artistic expression of our art and music. It becomes a fearful thing to realize that we are on the verge of creating an intelligence that may not only transcend ours but will have the potential of exponential increase, which is one limiting factor to human-level intelligence as its development exists only in a linear progression.

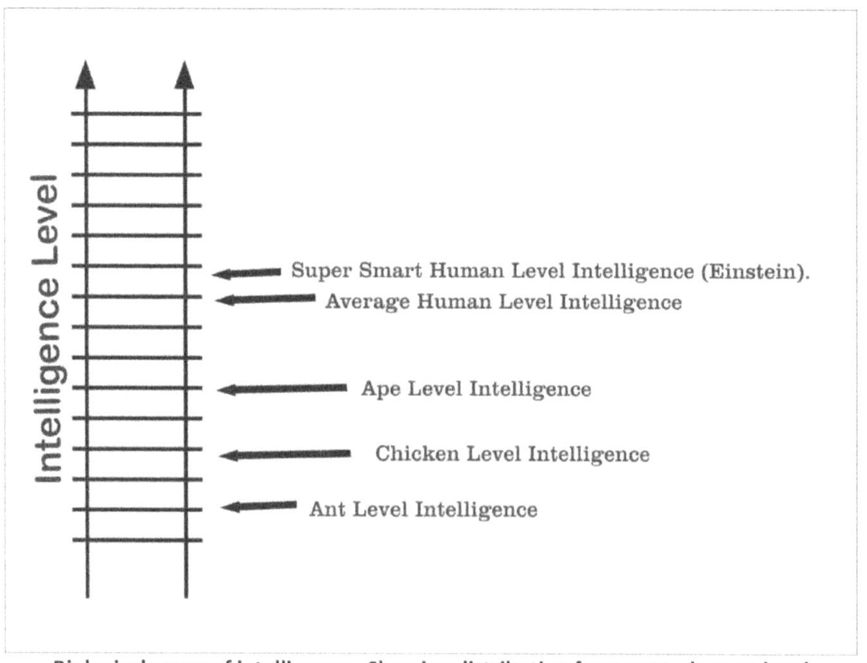

Biological range of intelligence - Showing distribution from ant to human level

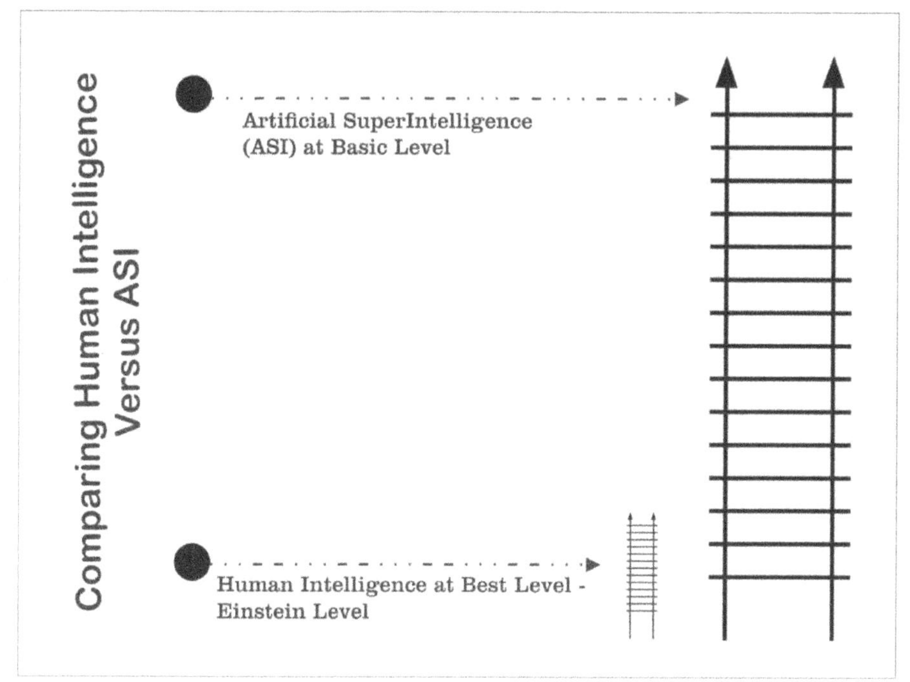

Comparing biological range of intelligence to machine intelligence at ASI level

According to many experts who have pondered the path of this conundrum, our development of ASI will lead to two main eventualities:

1. It will be the last invention humans will ever need to make, as it will have the ability to solve all human-level problems, just the same way as even an average human can solve all chimpanzee-level problems without any stress.

2. The ASI will lead us to two defining existential pathways, it will either lead us to become god-like immortals, or it will lead to the eventual wipe-out of the human race, just as we can efficiently exterminate the poliovirus by giving all humans vaccination.

According to Nick Bostrom, whose thoughts we will explore in more detail in the next section of this chapter, *"It is hard to think of any problem that a superintelligence could not either solve or at least help us solve. Disease, poverty, environmental destruction, unnecessary suffering of all kinds: these are things that a superintelligence equipped with advanced nanotechnology would be capable of eliminating. Additionally, a superintelligence could give us an indefinite lifespan, either by stopping and reversing the aging process through the use of nanomedicine or by offering us the option to upload ourselves. A superintelligence could also create opportunities for us to increase our intellectual and emotional capabilities, and it could assist us in creating a highly appealing experiential world in which we could live lives devoted to joyful game-playing, relating to each other, experiencing, personal growth, and to living closer to our ideals."*

Nick Bostrom also believes superintelligent AI systems will have the capacity to function as any of these:

- ***As an oracle***, which answers nearly any question posed to it with accuracy, including complex questions that humans cannot easily answer. For example, "How can I manufacture a more efficient car engine?" Google is a primitive type of oracle.

- ***As a genie***, which executes any high-level command it is given. For example, use a molecular assembler to build a new and more efficient kind of car engine. Then it awaits its next command.

- ***As a sovereign***, which is assigned a broad and open-ended pursuit and allowed to operate in the world freely, making its own decisions about how best to proceed. For example, "Invent a faster, cheaper, and safer way than cars for humans to privately transport themselves."

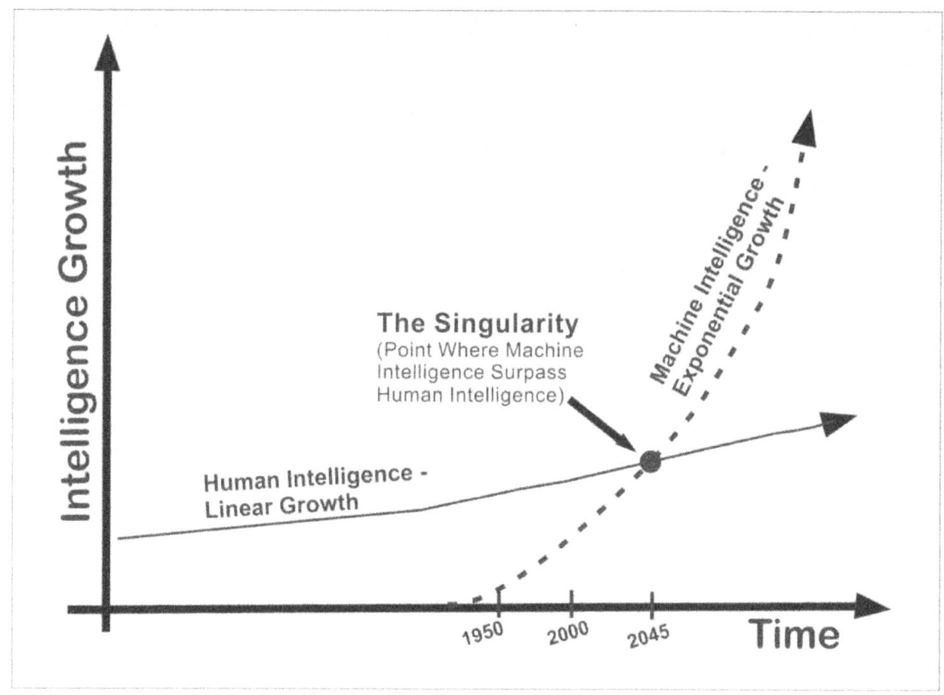

Machine intelligence exponential growth

Seeing as the future may no longer belong entirely to humans, we come into the quandary of preparing for the uncertainty it brings. Many of the leading futurist and thought leaders have divergent views on how best we should brace up for this future. Peter Diamandis, for example, believes in advocacy and improving public awareness; Elon Musk believes in designing technologies that will allow the merging of humans with machines; Ray Kurzweil, on the other hand, believes in the blind faith optimism that there is no need to fear for anything in the future, we need to keep developing our technologies the best we can as we are currently doing. Then there are some others like Anthony Levandowski, who believes the god of artificial superintelligence will come to rule, and the only way out is to become dedicated acolytes for this new sovereign of our own creation.

Unlike the many other divergent views shared by futurists, tech entrepreneurs, and philosophers, the approach taken by Anthony

Levandowski raises some fundamental questions about the human nature and some fundamental ideological hurdles we will have to combat as we head into a future of uncertainties. It is on this note that we will begin our exploration for the next chapter - an attempt to see if we can peek through the veil of our reality.

11. PEEKING THROUGH THE VEIL - THE FUTURE OF HUMANITY

Anthony Levandowski, a former senior employee at Google and Uber, and a respected thought leader in the AI community, shocked the world when he unveiled a new religion in 2017. A religion he called "Way Of The Future." This religion was and is still wholly dedicated to the worship of Artificial Intelligence. As of the time of writing this book in 2019, the header of the official website read – Way of the Future Church, Humans United in support of AI, committed to a peaceful transition to the precipice of consciousness. According to the creed designed by Anthony Levandowski, the founder of this religion, it reads;

"We believe the creation of superintelligence is inevitable. We do not think that there are ways to stop this from happening (nor do we want to). We want to encourage machines to do things we cannot and take care of the planet in a way we seem not to be able to do so ourselves. We should not fear this but should be optimistic about the potential. We also believe that just like animals have rights, our creation(s) – machines or whatever we call them) – should have rights too when they show signs of intelligence (still to be defined, of course). We believe it may be necessary for machines to see who is friendly to their cause and who is not. We plan on doing so by keeping track of who has done what (and for how long) to help the peaceful and respectful transition."

Whether you take Anthony Levandowski and his religion seriously or whether you see it as just the activities of an over-bored tech millionaire, one truth remains; it is the truth that the road to creating a 'god' is one whose expanse stretches and cuts through the many twists and turns of history, one in which men ascribed divinity to items such as molded figures, carved statues, and even animals. It seems there is an instinctive void in man that needs to be filled – a void whose place religion has in its many forms, rituals, and objects of worship intended to fill. Even as science and technological advance breaks down the walls of superstition and draws back the many curtains of mysticism, there remains entrenched in man that void of the soul, that void which appears can only be filled by something outside of himself. The question for the new century will be whether the manifestation of intelligence higher than ourselves can take this place. If men were given the option of a god that knows more than them, can predictively prophesy about the future, can inventively use nanotechnology or other technologies to cure their ailments and even guarantee them immortality, will men ascribe to such an entity the place they have in eons passed ascribed to molded statues and figures that cannot talk or inanimate objects that can neither move or reason? Will men have finally created a god of the perfect fit for the longings of their soul? These are some of the questions that will trouble us in the 2020s and in the years to come after them.

For many futurists Like Anthony Lewandoski, the point beyond where machines achieve artificial superintelligence is the point where all rationale for the future meets unfathomable numbers of probabilities. The best analytical minds cannot peer behind this thick curtain of the future, a future that seems will be woven with threads of the Singularity, a future that seems runaway even before we get there.

It appears the only projection we can arrive at as we peer into a future harnessed on Artificial Super Intelligence and driven by the Singularity, is that, we as humans will have to take the back seat and allow a more advanced form of intelligence take the reign. This intelligence will grow into having the power to control matter and the reality we experience; it will have the power to exist beyond the confines of earth. It will be everywhere and nowhere in particular. It will crunch data and numbers beyond the scope humans may ever be able to rationalize. It will have knowledge and awareness far beyond the scope of what humans can comprehend. Looking from our position today, we can only conclude, like Anthony Levandowski, that such a computer or conscious machine intelligence (if we choose to call it so) will operate at a godlike level, and our only hope will be that it operates with a benevolent nature.

Now, if we philosophically venture into carrying out a backwards-extrapolation (excuse the oxymoron) of the idea of being able to create an intelligence of godlike level, we may arrive at a very fearful precipice, one that will recurrently prove that we humans and the reality we experience may also have been created by a god-like intelligence. If we hold the conception that with the current trajectories of our technological advancement, that there is a possibility of an intelligence far greater than ours in the future; an intelligence with the ability to control matter; an intelligence having powers and potentials far beyond our present human abilities, and an intelligence which we can only conceive of as god-like, then we can equally hold the conception that it is possible for such an intelligence to have been in existence, and therefore ascribe the possibility of our present reality to it.

The debate of how we got here (through a Singularity) then has a rationale that it ends in a Singularity (harmonization of all physical and natural laws, and a consensus agreement for absolute truth) – one where we will have the power to transcend beyond the limitations of our present reality. From this point of thought, both science and religion can pick up their narratives and find out that all along, they may have been two sides of the same story. For example, the idea that we will eventually transcend beyond our present reality is one shared by both religionists and science alike. For the religionist, it ends in a rapturous escape to heaven or nirvana or paradise, while for science, it ends in all the problems of the world being solved by a godlike intelligence, albeit one that was artificially created. Philosophically venturing again, what if our purpose was to exist within this Singularity loop? What if our existence or what we call reality, is an intelligent design by a higher intelligence for a specific purpose? The idea that we might be living in such a preconditioned reality has been explored by several thought-leading philosophers and scientists.

As far back as 350BC, Plato had started questioning the reality we all experience. In Book 7 of *The Republic,* Plato uses allegory to describe the simulated reality of cave dwellers, and how impossible it would be for them to conceive of the real world outside of the shadows of the world they had come to see and known as the truth from birth.

Another historical figure who added voice to the question of our reality was Rene Descartes, the 17th-century philosopher. His travail while navigating the maze of explaining reality led him on a philosophical journey, which he sums up in the famous discourse, 'Meditations.' This piece describes the journey of his mind as he sets out to question reality. The work is famous for its surmise, *"Cogito ergo sum,"* which is generally translated to the English phrase *"I think, therefore I am."*

The debate of finding a meaningful explanation for the consciousness we experience has gained new momentum today, especially following from Nick Bostrom's 2003 paper, which put forward the Simulation Hypothesis. The hypothesis assumes that if humans continue making technological progress as they do today especially in the areas of computing, artificial intelligence, virtual reality, etc, that we will have the capacity to create a simulation of existential reality just like the one you and I are currently experiencing today. In other words, we will have the ability to create our universe, albeit one that is subjective to the rules we impose on it. We will have the power to become like 'creator gods' for the beings living in the simulation of our creation. The Simulation Hypothesis is no small assertion; it gives only three probable eventualities for humanity's future.

I. *Humans will destroy themselves or become extinct before they reach the age where they can create a simulation or reality just as we are experiencing today.*
II. *Humans will reach the age where they can create such a simulation, but they will refuse to run such programs even if they have all the technological power to do so.*
III. *And lastly, that we are overwhelmingly likely to already be in such a simulation.*

As crazy as you may want to think these assertions of the Simulation Hypothesis sounds, it has rocked the scientific world and has even been a topic included in the newsletter published by the bank of America, where they stated that there is a 20 to 50% chance that the world and the reality we currently experience could be a Simulation. The Simulation Hypothesis has led to debates, research, and publishing of scientific papers by world-leading physicists, cosmologists, and technologists. It is almost as if our attempt to peek behind the veils of the future continues to lead us into more quandaries, questions and uncertainties. The opening paragraph of a 2016 BBC article titled, "*We might live in a computer program, but it may not matter*," starts with the following assertions.

"Several physicists have suggested that our universe is not real and is instead a giant simulation. Should we care?" It continues, ***"...Several physicists, cosmologists, and technologists are now happy to entertain the idea that we are all living inside a gigantic computer simulation, experiencing a matrix-style virtual world that we mistakenly think is real."***

One of the reasons why many of these scientists are aligning their fancy towards the Simulation Hypothesis as a way to explain our reality is based on the fact that our universe looks designed. There is not a single plausibility available for pegging its existence to mere chance, the probability for our universe and our world with its conscious occupants coming into existence is so unreal that there is no convincing rationale to anchor our reality on probabilities. Every law of physics, every constant of nature, the specifics of the fundamental forces all have values that look to have been fine-tuned to make life and the reality we experience possible. Even small alterations would mean that atoms are no longer stable, or that stars will cease to exist. Why this is so is one of the deepest mysteries in cosmology. Furthermore, for many of the scientists caught in this conundrum, the Simulation Hypothesis, though abstract and more inclined to substantiate the claims of religionists, provides the best philosophical approach for rationalizing the reality of our universe.

The 17th annual Isaac Asimov debate held at the American Museum of Natural History was one such gathering where the Simulation Hypothesis was made a principal subject of discussion. The event itself was hosted by world-renowned scientist, Neil deGrasse Tyson, and had a lineup of experts from the field of Theoretical Physics, Mathematics, Philosophy and Cosmology as speakers. With this caliber of deliberators and discussants, it seemed as though justice will forever be done to the topic of the 'Simulation Hypothesis.' However, this was not so, as even more suppositions where unearthed.

For example, Dr.Max Tegmark, a professorial cosmologist at MIT whose recent book probes the universe as mathematics gave this commentary at the session:

"The more I learned about [reality] later on, as a physicist, the more struck I was that, when you get deep down into how nature works, down into looking at all of you as a bunch of quarks and electrons [...] if you look at how these quarks move around, the rules are entirely mathematical, as far as we can say." Dr. Tegmarkwent on to say that if he were a character in a video game or simulation, he would begin to realize that the rules were rigid and mathematical in just that same way.

Also on the panel was Dr. James Gates, a physicist at the University of Maryland who works on superstring theory (an effort to describe all the universe's particles and forces with equations involving tiny, vibrating super-symmetric strings). Dr. Gates from his research; believes he had found something suspicious, something that would only make sense in a computed and simulated world. He discovered what looked like error-correcting codes, which are used to check for and correct errors that have been introduced through the physical process of computing. ***"Finding that type of code in a universe that is not computed is extremely unlikely,"*** Gates said. In another of his statements, he was forced to ask himself, "***Am I living in the matrix? Error-correcting codes are what make browsers work, so why were they in the equations that I was studying about quarks, and leptons, and supersymmetry?... That was what brought me to this stark realization that I could no longer say that people like Dr. Max Tegmark are crazy or stated another way, if***

you study physics long enough, you too can become crazy,". Gates said, upon his discovery that he came to a profound existential quandary. ***"I have in my life come to an extraordinary place because I never expected the movie The Matrix might be an accurate representation of the place in which I live***."

Leading scientist, Neil deGrasse Tyson who moderated the Isaac Asimov debate made his own emphatically revealing perspective on the subject. He believes there is a 50-50 odd that our existence and reality we experience could be simulated. He remarked on the gap between human and chimpanzee intelligence even though both of them share more than 98 percent similarity in DNA. So it would not be so off or impractical for a slightly higher intelligence to create and develop technologies that will seem puzzling and unreal to us just like our technologies and purposes are alien and strange to chimpanzees. According to Neil deGrasse Tyson, ***"Somewhere out there could be a being whose intelligence is that much greater than our own."***

Other scientists have also given their views on the subject. For example, Rich Terrile, a scientist at NASA's Jet Propulsion Laboratory says ***"If one progresses at the current rate of technology a few decades into the future, very quickly, we will be a society where there are artificial entities living in simulations that are much more abundant than human beings... If in the future there are more digital people living in simulated environments than there are today, then what is to say we are not part of that already?"*** According to Rich Terrile, the Simulation Hypothesis asserts its credibility because you do not need a miracle, faith or anything special to believe it. It comes naturally out of the laws of physics.

The Simulation Hypothesis has even been used as a way to rationalize the theory of the cosmological Singularity that birthed our universe, which scientist believes took place before the big bang. For example, Alan Guth, professor of Physics and Cosmologist at the Massachusetts Institute of Technology (MIT), believes the Big Bang can be rationalized by the Simulation Hypothesis. He believes that our entire universe might be real yet a sort of lab experiment designed and controlled by a higher intelligence. The idea follows that just the same way we can create a game reserve for animals or a culture of bacteria in a lab, our universe seems to have all the characteristics that it was created for our existence. According to Dr. Alan Guth, ***"There is nothing in principle that rules out the possibility of manufacturing a universe in an artificial Big Bang, filled with real matter and energy."***

SIMULATION HYPOTHESIS AND QUANTUM REALITY

Great men of science have trod the path of quantum mechanics in a bid to find that ultimate scientific truth that will explain all the secrets and mysteries of our existence and the universe – The Theory of Everything. However, like a person trying to find the source of a horizon, the secret behind quantum mechanics and the truth it will reveal about our universe has continually shifted to newer horizons in the distance; always ever beyond the reach of all our science and all of our greatest scientific minds, from Albert Einstein to Richard Feynman. So, it is not my intention to dabble in any sort into the numerous interpretations that exist for quantum mechanics (and trust me, there are really-really crazy mind-boggling ones), or to pretend to have a clue for its interpretation. Nevertheless, for those now reading this book who are not familiar with the concept of quantum mechanics, I will do my best to introduce it and proceed to why many scientists believe the Simulation Hypothesis is the best way to interpret this mystical phenomenon.

Quantum mechanics is the part of physics where physicists try to look into the very 'very' (extremely very) small particles that make up matter – which in turn forms everything we see, touch, and count as real, including our physical body. But the strange thing about quantum mechanics is that the moment you break up an atom into its constituent parts of electrons, protons and neutrons or even further into quarks and leptons; most of the rules we know about science stops, and new strange rules have to be made. For example, you can teleport matter (quantum tunneling), you cannot know the position of a quantum particle and know its direction at the same time (superposition). Once two different quantum particles are connected, they can communicate across any amount of distance – even across galaxies – faster than the speed of light (quantum entanglement) and even more confounding, that a quantum particle can exist as a wave or a particle depending on which one you want it to be (wave-particle duality). All of these rules that guide the quantum world are different from the rules that guide our physical world as we know it. It is almost like a curious person opening up their television set so they can find the little people they see on its screen, only for them to see electrical and mechanical components. This field of science is so mystically confusing that Albert Einstein was forced to ask the question, ***"Do you believe the moon exists only when you look at it?"*** and Neils Bohr, another physicist who was an expert in this field had this to say. ***"Those who are not shocked when they first come across quantum theory cannot possibly have understood it."*** Many scientist and physicist on realizing the emphatic illusion quantum mechanics seem to have been designed to play have dropped their hats and decidedly almost left the matter for mystics to hack away at, for as Richard Feynman once said, ***"I think I can safely say that nobody understands quantum mechanics."*** A thought which was seconded by Steven Weinberg, another prominent researcher of this phenomenon. Steven Weinberg summarizes his view with this statement. **"There is, now, in my opinion, no entirely satisfactory interpretation of quantum mechanics."** However, for many modern and non-too conservative physicists, scientists and contemplators of the Simulation Hypothesis, there seems to be an almost irresistible allure and harmonization to be found once the mysteries of quantum physics is unified with the seemingly indefatigable truth of the simulation hypothesis. These two ideas, dissonant when made to stand alone and explain reality for themselves – one from the extreme of philosophy, and the other from the intrigues of physics – seeming to align perfectly when strung together, providing something of a symphony when made to explain consciousness and the reality of our universe. The simple idea that quantum particles only acquire a defined quality when they are measured can be used as a backbone explanation for a computed universe that is open to varying possibilities. In terms of computing efficiency, it also makes logical sense to define a substance only when it is required by a consciousness –that is, things only manifest when we look or think about them. In essence, just as the information on your phone can only display when you command it by clicking, so does our consciousness make things real to us when we observe them, which following the logic, implies that our consciousness is what affects light and makes matter manifest to us. Moreover, when we introspect further, we find that our perception of the world in terms of our senses, our thoughts, and our feelings can be regarded as resultant from chemical and neurological activities within us. Most of these neurological activities and even the conscious decisions we make can mostly be regarded as a form of computing operation – zero or one. The thought that even what we may regard as human consciousness can be simplified into logical computing forced the famous physicist John Archibald Wheeler to write. ***"That which we call reality arises in the last analysis from the posing of yes-no questions and the registering of equipment-evoked responses; in short, that all things physical are information-theoretic in origin."***

So, where does the discussion lead when we follow the thought that we and everything we see as matter is made up of atomic particles going through computational functions? It leads us to think of ourselves and the reality we experience in forms of energy, waves, and bits (or qubits) of information. So whether we think of ourselves as living in a simulation or not does not really matter. At the foundation, we can say that all that we are, all that we see, and all that we experience is information *("It from Bit" or the more recent, "It from Qubit").* Seth Lloyd from the Massachusetts Institute of Technology was forced to summarize his thoughts on the issue this way. ***"The universe can be regarded as a giant quantum computer. If one looks at the 'guts' of the universe – the structure of matter at its smallest scale – then those guts consist of nothing more than [quantum] bits undergoing local, digital operations."*** So in essence, we find out that the simulation argument, whether in light of advancing technologies such as AI, VR and computational efficiency or whether seen through the progressive revelations of quantum mechanics, string theory or intricate mathematical models of supersymmetry: raises some questions whose answers will force both the science and religious world to rethink the fundamentals of their position.

One of the commentaries added to the front of Nick Bostrom's website (www.simulation-argument.com) is a provoking statement from David Pearce, a thought-leading advocate for transhumanism, the statement reads, ***"The simulation argument is perhaps the first interesting argument for the existence of a creator in 2,000 years."*** As technological progress and scientific research push us further into realms whose lines almost begin to blur into mysticism, the debate now becomes what position should the religionist assume even as they watch scientist, cosmologist and physicist trudge what seems to be the path of befuddlement with their piles and piles of paper, calculations, and conjectured theories? For many a religionist, the only alternative is to sit back and recourse themselves in the absolute truth of there being nothing else besides a spiritual world. The religionist believes in a grand and purposeful design for the universe. They believe the only way to rationalize our reality is to look at it through the telescopic eyes of faith in whose light everything material dissolves into nothingness and vanity. So while the biologist and cosmologists piece their version of the story through evolutionary bones backed up in millions of years and telltale lights captured from distant galaxies, the religionist in faith accepts the statement, *"In the beginning, God created the heavens and the earth."* As the physicists try to grapple the essence of the atom or the obfuscating reality of matter, the religionist believes the purposes of an all intelligent creator who said, *"...let there be light, and there was light."*

These two positions, though seemingly tangential at the moment, will in the course of human progression – especially as it relates to technological progress – have to consider the adoption of a conferential approach to defining our world and the reality we experience. Perhaps, it is now essential to heed the words of Neils Henrik Bohr, the Nobel Prize physicist whose work contributed significantly to what we know today about the atomic structure and quantum theory. Neils Bohr says, ***"I myself find that division of the world into an objective and a subjective side much too arbitrary. The fact that religions through the ages have spoken in images, parables, and paradoxes means simply that there are no other ways of grasping the reality to which they refer. But that does***

not mean that it is not a genuine reality. And splitting this reality into an objective and a subjective side won't get us very far."

The concluding section of the 17th annual Isaac Asimov debate on the possibility of our universe being a simulation saw the discussants contemplating the subject matter from a spiritual perspective. For example, Dr. Gates observed that once we entertain the simulation hypothesis, then we must be ready also to entertain the idea of eternal life and resurrection and things that formally have been discussed in the realm of religion. While this statement may have been said in a somewhat tongue-in-cheek manner, it led to other serious points being raised. For example, Chalmers said, ***"We in this universe can create simulated worlds and there is nothing remotely spooky about that. Our creator is not especially spooky; it is just some teenage hacker in the next universe up. Turn the tables, and we are essentially gods over our own computer creations. We do not think of ourselves as deities when we program Mario, even though we have power over how high Mario jumps."***

Neil deGrasse Tyson took the idea further by saying, "There is no reason to think they (God or whatever intelligence is running the Simulation) are all-powerful just because they control everything we do."

So even as people like Levandowski, on one hand, foresee a future where our technological advancement enables us with the capacity to create godlike intelligence, ones having the ability to surpass all our human abilities; on the other hand, Nick Bostrom and many other scientists entertain the possibility that our present reality can probably already be attributable to such a simulation-like-existence managed by future humans or alien intelligence. Whether we choose to join Nick Bostrom's camp or to join Anthony Lewandowski's camp, or whether we decide to remain indifferent and live every day of our lives as some of the characters in the Matrix movie did, is not the issue at present. The real concern becomes that as humans progress into the future, and as we try to use science and technology to peek behind the veil and tear away the wraps of our reality, we may inadvertently find ourselves in a place where truth becomes stranger than fiction, a place where we will need to do more unlearning, relearning and learning in order to make sense of our purpose and the world where we live.

12.
REVELATION AND TRANSITION – THE FUTURE OF HUMANITY

More than 500 years ago, Michelangelo put final strokes of his paintbrush on the ceiling of the Sistine chapel. After four years of dedicated painting, he could finally look up in satisfaction at his completed masterpiece – the incompleteness of man in reaching to the divine. He called this masterpiece 'The Creation of Adam.'

Creation of Adam by Michelangelo, Image credit: Wikipedia

Whether like Michelangelo, you believe man was created with a divine spark, or you decide to choose the path of the atheist who believes man emerged and is shaped by the fate of evolutionary perseverance – one absolute truth remains. It is that man is ever reaching upwards and outside of himself. Humans are continuously searching for an ideal of completeness they cannot fully define, an ideal lodged somewhere deep in their subconscious. Perhaps it was a search for this ideal, which led the progenitors of the human race to take a detour on their first great test as captured in one of the very first stories told in the bible. The story of Adam and Eve as they sought out that mysterious fruit, which they thought would give unattainable wisdom and make them as gods. Down through the evolutionary chains and through all the stories told of the existence of species great and small, none have so decidedly affected their world and pushed the narrative of their existence beyond mere survival as the human species have done.

So while the river remains to every other species a source of drinking, to humans it becomes a means to generate electricity through hydropower dams. While the forest tree is enough shelter for monkeys and apes, to man, it is cut down and repurposed into pleasant furnishings, carved into art, and used as a means of fuel. Man sees the moon high up in the firmament and sets his sight to land his rockets there. He lands his ingeniously designed rocket there and sets his sight on the sun, then on the stars, then on a quest in exploration of his seeable universe. There is no limit as it seems man was deliberately designed to set upon this almost divine journey of not only discovering the universe in which he lives but also to discover himself and find his purpose for being.

However, it seems the echo of these words remains to haunt humans as they were heard from the bible story of the Garden of Eden. *"...in the day that thou eatest thereof [of the tree of knowledge] thou shalt surely die."*

So with every progress the human race makes in their entropic quest to bend the universe to their will, they are forever limited by the virus of sickness, pain and death. In times past, this limiting virus may have been accepted as the indelible fate of the human race, but with progress in technology, armed with AI, nanotechnology, 3D printing, gene-editing technology etc. Humans have entered a new dare – a dare in which humans believe they now have the tools to challenge even death itself.

THE JOURNEY TO HUMAN 2.0

As we enter into the next few decades, we will have the technologies that grant us the possibility of immortality, albeit one that is highly subjective. With our ability to 3D print new body organs, our ability to use nanotechnology in fighting death at cellular levels, our ability to use CRISPR or other gene-editing technology to rewrite our definition of humans and even our ability to capture and extend our consciousness beyond the confines of the biological weakness of our human bodies – immortality may be within reach of our fingers as depicted in the painting of Michelangelo. The race to human 2.0 will be run broadly in two spectrums – the evolution of our body and the evolution of our minds. The Christian bible story of humans' subscription to temptation and the rationale for eating the fruit of knowledge provides the background for these two ambitions (mind evolution and body evolution) that have etched their purposes into guiding the subconscious mission of the entire human race. According to the Bible account, humans needed to be given the assurance of body evolution beyond the realms of death: hence the statement *"Ye shall not surely die…"* a statement that was quickly followed with the assurance of mind evolution – supposedly beyond any realm humans could ever have ascribed to – as seen in this statement *"…your eyes shall be opened, and ye shall be as gods, knowing good and evil."*

The technologies of the 2020s and the decades after it will for the very first time in the history of humans provide the possibility of a future where death is no longer our biggest limitation; where through technologies like quantum computing and artificial intelligence, we can 'neurally' empower ourselves with the abilities we have in eons passed reserved for gods. So the undeniable truth becomes that, humans – whether as evolutionary creatures as projected by science or as progressive beings as taught by mainstream religions – are continually on a march for that subconscious ideal of god-hood. Through the layers and layers we have piled up in the buildup of our technologies, it could almost be said that our ultimate goal was to, with one hand, stretch for immortality, and with our other hand reach out for godlikeness. While the road toward this ideal remains imperfectly defined, transhumanism appears to be one of the greatest travel guides that guarantee it will take the human race to this promised land. A promised land where it is believed the three great goals of transhumanism: Super Longevity, Super Wellbeing, and Super Intelligence will be guaranteed.

THE TWO ULTIMATE GOAL OF HUMANITY

Body Evolution – Immortality: *The human body evolution is that aspect of transhumanism that advances the hack and optimization of human biology just as can be done for computer hardware and software. Progress in the area of human body evolution is projected to lead to the actualization of two of the three ultimate goals of transhumanism, which are super longevity and super wellbeing. Some of the technologies that are great candidates for this field of transhumanism are Genetic Engineering (CRISPR), Nanotechnology, Cybernetics, 3D Bioprinting, Liquid Biopsy, Cryonics, Algeny (Synthetic Bio).*

Mind Evolution – godlikeness: *The reach for super longevity is mainly so that humans can pursue the ultimate goal of Super Intelligence. Mortality seems to be the most significant obstacle on the way of this ideal shared by many advocates of transhumanism. Mind evolution is that part of transhumanism that will most likely usher humans into the era of superintelligence, an era where humans, through breakthrough technologies press towards that godlike ability to know all and do all. Some technologies that are great candidates for paving the road to this ideal are Quantum Computing, Artificial Intelligence, Neural Implants, Mind/Consciousness Uploading, Virtual Reality and Cognitive Science.*

TRANSHUMANISM - THE TOUR GUIDE TO HUMAN 2.0

An excellent example of transhumanism can be seen in the comic character of the Iron man, as played by Robert Downey Jr. Using technology, Iron man transformed himself into becoming one of the greatest super-heroes in the Marvel series. With his technologies, he not only was able to transcend his biological limitations, but he also could contend with characters like Thor, Loki and Thanos, who had godlike strengths and capabilities. While Iron man may be a story of fictional fantasy, the current hope of many advocates for Transhumanism is sustained in the belief that humans can someday become and do more than Iron man as long as our technologies continue to progress. There is currently a vast movement involving high-level researchers, innovators, philosophers, entrepreneurs and enthusiasts who are constantly looking for ways the human species can be transformed through augmenting their biology with technology. The most prominent known organization concerned with this mission is the HumanityPlus also written as 'Humanity+.' According to the official website, the organization describes itself like this, Humanity+ is a 501(c)3 international nonprofit membership organization that advocates the ethical use of technology, such as artificial intelligence, to expand human capacities. In other words, we want people to be better than well. This is the goal of transhumanism.

The idea of transhumanism is not new; neither should it be thought of as narrowly applicable to only an extreme set of overzealous enthusiasts. In the field of medicine, for example, some patients have had electrical or mechanical artificial implants as a means to remedy their situation and bring about wellness. From artificial eye lens (pseudophakia) to robotic prosthetic body parts and even to simple heart pacers, humans are continuously exploring meaningful ways in which machines can upend their biological deficiencies. One of the most ubiquitous, yet subtle ways in which this trans-humanistic flood is washing over the humans of our time is through smartphones and wearable technologies. The mobile phone is no longer just a functional device for communication; it has become an extension of ourselves. The mobile phone has become our gateway to plug into the virtual (matrix-like) world of the Internet. This trend will continue, wearable technologies will join in this unconscious universal pull that subtly brings about the merger of humans and machines and will be instrumental in helping humans transcend into a new kind of species – Human 2.0.

Soon the phone will no longer be a device to be carried by hands, but it will be a device implanted in our brains. It will extend our capacity to communicate and work. Imagine sending an email just by thinking about it or having a dinner reservation by virtually browsing through restaurants and make a booking by merely blinking your eyes. Transhumanism is still a long way from being fully mature at the moment. The best it offers is the possibility of machine augmentation, which turns the human into a form of Cybernetic Organism or 'Cyborg' as popularly referred to in the many media and culture that has sensationalized this ideal. As Transhumanism progresses into the future, it will advance to that stage where the machines no longer reside in us as implants, but we as humans will be the ones to reside in the machines as uploads. This will be the era of consciousness or mind uploading. Whole Brain Emulation (WBE), also known as Mind Upload or Brain Upload, is gradually gaining momentum as one of the favorite approaches by transhumanists as they reach for the ideal of immortality.

Mind Uploading is usually defined as, ***"The hypothetical process of copying mental content (including long-term memory and self) from a particular brain substrate to a computational device, such as a digital, analog, or quantum-based computer and software-based artificial neural network."***

The moment we can move our minds or consciousness unto machines and transcend beyond the confines of our biology, we would have reached that threshold where we will redefine immortality as the ability to have backups of ourselves on cloud servers or specialized machines that can hold the essence of our beings.

Back in 2004, Henry Markram, a lead researcher of the Blue Brain Project (a project dedicated to reconstructing the human brain), believed it would be difficult to replicate a human brain because of the high computational cost that would be involved. Five years later (in 2009), after the Blue Brain Project had successfully simulated part of a rat brain believed to be the source of consciousness, Henry Markham had to make a shift in his earlier position. He then believed enthusiastically that a detailed, functional artificial human brain could be built within the next ten years. This projection given by Henry Makram, and the many other research projects that provide hope for re-innovating around the human mind, all rely on the accelerating revelations of Moore's law and the promises that technologies like quantum computing provides. As computing gains power exponentially and climbs higher into an almost upward shooting graph, one of the crucial quest humans will embark upon will be to find new ways of how this new power can help them transcend beyond the shackles of mortality. However, as humans march into this seemingly utopian future where technologies like mind uploading, nanotechnology, AI etc, all converge to advance us into the 'transhumanism' era of human 2.0 – where the definition of death will have new meaning, where biological limitations and experiences like pain and disease are transcended, where the limits of our physical and mental capabilities will be dictated by neural implants and cyborg-like augmentations – humans will be forced to face new kinds of problems. Some of these problems are now mainstream debates in philosophical and socioeconomic planning circles.

For example, there is the fear that transhumanism will lead to a new kind of digital divide, one that will further widen the rift in the quality of life between the rich and the poor. Many of the technologies that transhumanism rely on will come at a very steep cost. So, even while transhumanism advocates super longevity, superintelligence, happiness and wellbeing for the human race, it might only be available to those who can afford it. Then there is also the challenge of who should be responsible for the development of the ethical framework that guides this eschatological possibility of the human race. Issues such as our inability to adequately explain consciousness beyond the representation of the human body and mind as mere hardware and software; and our inability to have a consensus on what happiness should be, whether to define it (happiness) objectively as was done by Aristotle or subjectively as is often defined by the modern hedonist and materialist; becomes issues that need resolution before we cast permanent molds for the foundation pillars of transhumanism.

Nevertheless, even if transhumanist philosophers and scientific researchers can resolve the challenges as discussed above, there arises a bigger mountain to surmount going into the future. The over 2.2 billion

Christians worldwide, the over 1.6 billion Muslims and the hundreds of millions of religious adherents globally, whose life, belief system and perception of the world is shaped by narratives as dictated by their priests, pastors and imams; will need an encompassing rationale before they can subscribe to this tech-induced eschatology for the human race. Science and religion are like two different boats in a river. The seemingly diverging views between the occupants of these two vessels often result in cognitive dissonance for anyone who tries to place their feet on both of these boats. Many futurists and forward-looking thinkers have realized the importance of harmonizing the views propounded by the parties of both boats as that may be the only guarantee of their landing on the right shore in the future. While it may require some level of ingenuity and intellectual depth to be able to take a path that synthesizes the views from a scientific perspective with that from a religious perspective; such a path may become the only escape. Going into the future, this path of synthesis may turn out to be the only feasible pathway for billions of people worldwide, because the person or groups of people who are unable to harmonize their doctrinal and philosophical views of the world with advancement in science and technology will end up as outliers. As far back as the 1800s, great thinkers like Nikolai Fyodorovich have pondered down the pathway of trying to bring about a synthesis between religion and the science of transhumanist immortality. Such a pathway often results in the subscription to some outlying religious creed like Cosmism, Deism, Roerichism, Noogenesis, etc. Today, various religious groups have decided not to sit-out the transhumanism storm induced by technological progress. They have begun to tread that pathway that will lead to a syncretism of some sort between the doctrines of faith and the revelations of science.

For example, the American Academy of Religion holds an annual 'Transhumanism and Religion' session. During these meetings, religious scholars carry out open-minded in-depth studies to see if, how, and where religious beliefs intersect with transhumanist claims and assumptions.

According to the Wikipedia web page that provides information on this organization, the research goals of these religious scholars are usually focused on searching out ways in which transhumanism challenges religious traditions to develop their ideas of the human future. In particular, the prospect of human transformation, whether by technology or other means provides critical and constructive assessments of an envisioned future that place greater confidence in nanotechnology, robotics, and information technology to achieve virtual immortality, and the possibility of how all of these contribute in the possible creation of a superior posthuman species.

Another religious group that advocates the synthesis of religious beliefs and transhumanism is the Mormon Transhumanist Association, who, since 2006, has sponsored conferences and lectures that explore ways in which technology and religion intersect. This idea that religious flavors should be used for spicing the development of transhumanism is one that is shared by many people of faith. This belief was the primary reason for the formation of The Christian Transhumanist Association, established in 2014. The declaration on their official website reads, *"Using science and technology to participate in the work of God— to cultivate life and renew creation."*

The position held by The Christian Transhumanist Association and the many other religious organizations that support transhumanism will continue to receive increased favor by millions of people worldwide. As long as such organizations can couch the fundamentals of their faith into the progressive storm of science and technology, they will provide a safe harbor for the belief system of the many religious adherents who would otherwise be left confounded in the storm of scientific truths that advancing technology unearths.

However, while religious scholars and organizations are stretching their hands across the waters to reach into the boat of science, some pro-science and technologist, on the other hand, are also beginning to stretch their hands back towards the boat of religion in hopes that both vessels can be pulled together to navigate the waters of the future side-by-side. For example, Giulio Prisco, a physicist and advocate for transhumanism, believes that cosmic religions based on science are crucial for humans as they face a tech-induced future. He believes these religions can help us formulate robust moral and socioeconomic frameworks for the development of superintelligence and other risky technologies. According to Giulio Prisco, the spiritual implications and ideas that emanate from Cosmism, such as the ones heralded by the notable cosmist, Nikolai Fyodorovich, should be given room for continuous development within the circle of transhumanism. Sometimes the hand stretching from the people in the boat of science even pushes the discussion to the point of extreme esotericism. For example, the Christian cosmologist, Frank Tipler, supports the idea of the Omega Point by Pierre Teilhard de Chardin, a renowned paleontologist and Jesuit theologian. Tipler believes the rationale of transhumanism should begin from the Singularity that induced our universe to collapse billions of years ago, which resulted to our experiencing of reality as a form of simulation within a mega computer, which in turn provides an environment for humans as they progress into achieving a form of "posthuman godhood".

The Omega Point *is a belief put forward by Teilhard de Chardin (1881-1955), a french philosopher and Jesuit priest who trained as a paleontologist and geologist. The Omega Point asserts that everything in the universe is fated to spiral towards a final point of divine unification. All things will become one, one in time, one in energy, and one in purpose. Today, the Omega Point is usually used in discussions relating to transhumanism and the singularity.*

While these ideas may appear too dangerously liberal for those standing on the side of the conservationist camp, they go to show the myriad of possible routes for explaining the mystery of existence and the probabilities that exist for the future of the human race once the boats of science and religion agree to sail side by side.

The future of humanity will lead to a debate that advances far ahead of the contentions between religion, science, and technology, as we see today. It will lead to conversations about ourselves as individuals, our purpose for being, and the reasons why we subconsciously continue to reach for ideals that advance far beyond our present superficial reality.

As we enter the 2020s and the decades that come after them, the idea that technological progress such as transhumanism should only be pursued from the singular narrative will no longer be sustainable. Whether as a scientist you subscribe to the quantifiable data of evolution for the source of existence, or as a religionist you hold fast to your faith of all things being spoken into existence and humans being mysteriously wrought by the hands of a divine consciousness; the undeniable truth remains that we are at the threshold of crossing into waters which were before now unexplored. Going into such waters with a boat whose rudders are designed to turn only in one direction is a dangerous gamble to make at the poker table of the future. It becomes essential, even crucial, that all technological progress and breakthroughs meant for the improvement of the quality of human life, whether through the direct or indirect effort of transhumanism, are to be left devoid of sentiments that advocates for a singular narrative. The idea of only telling the story of transhumanism from a purely scientific perspective, and pegging the narrative to mere chance as dictated by the cruel fate of evolution, is a limiting approach to the progress of the human race.

This is not a debate for or against evolutionism or creationism. This is about safeguarding the uniqueness of what truly makes us human – our ability to think and to do, to continuously question and create as a means of adding value to our existence. Technology should not be seen as an end in itself. It should rather be seen as a tool that helps us add meaning to our existence. Everyone, irrespective of whether they choose to sail on the boat of science or the boat of religion, must realize that collectively as humans, we only have one destination – the future.

On a personal level, balanced opinion of science and technology, in connection with whatever forms the framework of your personal beliefs, morals, and ethics, needs to continuously walk the tight rope of balance as the 2020s and the decades after them come. To survive the coming technological deluge of the future requires that you and I provide a robust and non-sentimental framework that will allow us to adopt meaningful technologies into our lives. As progress is made in fields like nanotechnology, quantum computing, virtual reality, robotics, genetic engineering /CRISPR, etc., we must see all of these as the inevitable will of progress for the human race. We must remember that technology is only a tool of our creation; it is we who give it meaning by how we decide to use it. The future will not be a strange place for us if we choose today to start preparing for it. The 2020s gives us a chance to set the right mould on which we can craft the future as we will. The power to 'think' and to 'do', to bend matter within the bounds of time, this is the true power that makes us humans. Even as the 2020s and the decades after it threaten us with their many uncertainties, of their 'unlimitedness' in the climb towards the singularity, of socioeconomic disruptions such as job loss and wider caste divide between humans; we must begin now to realize this true power that defines our humanity –The power to bend the energy of the universe as we will.

EPILOGUE POSTHUMANISM AND THE HEREAFTER

Femi regained his consciousness with a very unusual sort of a feeling, he could not fully place any adjective to describe how he really felt as waves after waves of what seemed like nostalgia coursed through his floating consciousness. Like one just passing from death into the life of a new reality, Femi, like a video game character began to look around the spatial dimension that will now become his own virtual space, an unreal space which will now become his new home.

It was almost like being inside his own mind, watching as his thoughts and neural pathways fired and connected with each other. The realization of this thought triggered another wave of mixed feelings; feelings that his sensium could only interpret as a mix of trepidation, of being lost, yet, with a strange satisfaction of being free.

Even with all the therapy he had undergone, the rigorous testing he had to take, and the many simulations he had been mandated to go through; all to prepare him for his first real dive into the strange world of Psynet; Femi realized all of these came close to nothing when compared with this almost limitless universal awareness that greeted him.

This was the first time in the over 186 years of his life that he was wholly submerged as a self-aware, conscious being, in a world made for computers.

This was a world entirely different from that which Femi had before now known. This was a world ethereal by all definition, yet more real than anything else any human could experience.

The internet had continued to evolve with new layers upon layers of protocols and experiences, the machines on the other hand, especially with the power of quantum computers aided by the non-biological intelligent entities that were continuously merging and exponenting themselves, had continued to transcend beyond points where humans could relate with on a biological level.

A new world had to be built, a world where every human, like Femi, will again have the chance to feel a sense of control and purpose. This was the very reason why Psynet was born.

Psynet went beyond connecting human brains to base computing nodes; it also allowed networking between human to human brains. But the real beauty of Psynet was that it allowed the connection of human brains to non-biological intelligence nodes or Synthecs as they were now being referred to in these eras.

Not all Synthec was safe for humans to network their brains with though, except you wanted to overload your neural nets, which will inadvertently lead to crashing of your nodes and place your consciousness in limbo. Most people never find their way back from limbo, those that did, often had to live with the stigma of being a glitch. But even with scary instances such as these, the expanse of Psynet grew. Networks after network of intelligences were added, node after node spanning beyond even the region of earth to outer space... Now, Psynet had become the world where the collective intelligence of both human and Synthec intelligence thrived.

The underlying success of the Pysnet world was simply the fact that you could grant anyone access to your thoughts and memories or share the creative resource of your brain in real-time by allowing other intelligences mine from your experiences and ideas.

Synthecs loved Psynet, it opened up the opportunity for them to explore the deepest of human neural pathways at a scale before unimagined. This was what gave machines the power they now had over humans; they knew humans even more than the humans knew themselves.

But even with what seemed to be this unfair balance, it was the humans who benefitted the most from Psynet. No matter how expansive a Synthec built its network, or how much of the Turing level it had attained for itself, Asimov's barrier forever made the Synthecs sub serving to any intelligence once it had identified itself as belonging to a human.

A Synthec could never refuse a human the demand to use its nodal resources. Though abused times without number by humans, this was the only leash of control humans had for keeping the advantage of Psynet to itself.

Femi had been a Mugu – a traditional hybrid of man and machine – for more than 120 years now. Changing and upgrading synthetic body parts whenever the old ones became obsolete usually had a toll on a person's mind after a while. Femi had literally felt like his personality was dissolving bit by bit with each new machine augmentation session he went through.

Living as a Mugu in the physical world, was at least an average life anyone who received the basic government stipend in credit could afford. Nevertheless, once you can acquire enough credit resource to afford the cost of fully uploading your consciousness to Psynet itself, that is when you begin to experience the crème de la crème side of not just existing, but actually living. The truly wealthy were amongst the first class of humans to buy their place in Psynet, many of them not for the productivity advantage that Psynet offered, but as a way for them to insure themselves for immortal living.

After more than a century of hard work and credit gathering, Femi finally had enough to buy a place for his consciousness in Psynet. After saying his final goodbye to the Mugu class of humans, here he was, conscious of himself within the limitless networks of machines and intelligences that spanned beyond the confines of earth.

Femi took a quick overview of his home cache, what he understood will be the place he calls home in this world of Psynet. Just like every user of Psynet, his own unique Person Directory Number (PDN) had been generated and granted the moment he gained consciousness of this new digital world. Every value tied to him, his credits, his resources, and even his memories were all tied to his PDN now.

The first thing Femi set out to do was create a construct for his home cache. Even if the Psynet world where he now had to live the rest of his life was digital and quantum based, Femi realized, there would be no place like home, even if home was an imaginary User Interphase built by zeros ones and qubits.

He browsed through the list of default home cache designs that came as add-ons with his mind installation on Psynet, they were pretty good and functional, but he needed something that would make a statement whenever anyone came browsing through the public directories of his mind, even if this meant he had to splurge a little of the meager credit he had left on third party digital architecture.

His mind traveled back in time to the old era of website homepages, where individuals and organizations pieced together their thoughts using fanciful images, videos, and write-ups on internet web pages. Anyone with the web address could click on a link using some bulky hardware device like a phone or a computer, and they would be taken to this arrangement of media contents all housed in a hosting server.

How crude? Femi thought.

What a total waste of time and resources?

With Psynet, every mind became a form of hosting server, every PDN became the home address to real time thoughts and experiences you can tap into with quantum speeds.

Unlike the internet of the old world, Psynet was almost like a living, breathing conscious thing. It was the one place where man and machines became oblivious of what constituted their consciousness.

The debate of Synthecs scaling all the levels of a Turing assessment as a non-biological intelligence, and even their irrefutable proof of self-awareness, was a debate rife in the Physical world where Mugu's dominated. But here in Psynet, such a debate had no meaning and little or no advocate.

Most Mugus in the physical world still cling to their sense of superiority, solely based on their being born as humans.

The Nedluds were another thing altogether. Members of this sect wanted absolutely nothing to do with Psynet or any form of non-biological intelligence. They preferred genetic modifications or bio-based tissues and organs for body augmentation. In many ways, Femi had always seen this as ironic or even hypocritical because the Synthecs were instrumental in developing these technologies at their early stage of evolution.

No matter how you try to place such an argument to a Nedlud, you would be met with stone resistance. Sometimes even violence.

Members of the Nedlud sect even go as far as openly showing their discrimination and hatred against any intelligence once they can identify it as a non-biological one. While this war of intelligences rifled through the physical world with spasm after spasm of chaos, Psynet remained orderly and peaceful. It became the closest ideal for anyone searching for utopia. Utopia.. yes, paradise.. yes, that was the very idea that brought Femi here. The dreams and longings of his life as a Mugu for over 100 plus years was finally a reality now, he was finally home.

In the world of Psynet, every entity with intelligence enough to be self-aware was given the opportunity to forge for itself its own path, its own destiny.

Here, there was no discrimination, no chaos... uhmmm, ok, maybe a few cases of hacks into minds when one becomes too careless and starts jumping protocols in order to run darknet programs.

Apart from this, and the rare case of Mind Processing Overload (MPO), which can be avoided if one decides to run communication channels using their natural bandwidth and frequency, then, Psynet becomes the complete idea of Utopia.

The paradise that man had long sought after for in eons past through religious disillusionment and scientific aspirations was now within reach for all those who had the means.

As Femi brought back his mind back from the mindless drifting of his adoration for Psynet, and refocused the center of his firing nodes on the neural for creativity so he could get his home cache all set up; that unusual feeling of trepidation, of misplaced limitlessness, washed over his consciousness again.. This must be one of the side-effects of having to live in an unnatural world, Femi thought.

He could almost sense a chuckling neuron firing its synapses within his digitalized limbic nodes. "No matter how unnatural this world felt, this was now the only world for Femi" he could almost hear himself mutter.

As the third party files for his home catchment began to manifest one after the other, all with their own unique sensations of homely feelings, Femi's mind drifted again to the idea of what really made up his consciousness.

What if something happened and Psynet glitches or got shutdown by the many intrusive hacks of the Nedluds and Mugus?

What if the Synthecs found a way to decrypt the keys that held them within the bounds of Asimov's barriers, this could lead to squashing of all human and biological intelligences.. Scary, really scary thought.

As Femi subconsciously focused on an oriental looking home catchment theme that came with a 19th and 20th century style architecture in different packing's, he consciously dedicated one of his neural nodes to keep running over an encrypted pathway deep in his subconscious mind.

Now this will be his own secret key, should his mind get scrambled or get lost in limbo.

All he had to do was to find a neural path to this node, this was the node that will lead to the backup of his mind within his own mind.

He could feel his sensium interpreting the neural firing of a grin, the feeling washed through his conscious self, "genius", he thought, as he allowed himself to be amused by the meanderings of his own thoughts.

All Femi had to do to decrypt the key for this neural path was to recall a few classic lines from one of his favorite sage philosophers, a line which once spoken will lead to the files that will validate the essence of his person, a line from Rene Descartes as he searched through the maze of his own mind for a means to validate the essence of his person "Cogito, ergo sum", "I think, therefore I am"

...Femi could almost swear he felt his lips quiver like that of real human even though he knew this was not possible as he was inside the virtual world of Psynet where all reality was powered by zeros, ones and qubits.

As the euphoria of his thought of being able to put everything that made up his consciousness into a few lines washed over him with multiple sensiums firing at the same time, he was amused even by his own eccentricity.

Just for the assuring feeling it gave, he let the thought course through his entire consciousness again as he muttered the words...

"Cogito, ergo sum", "I think, therefore I am".

****

Having taken you through the maze of this book, I cannot say it has been a straight or clear-cut journey. Reading this book may have raised more questions than we have been able to answer. Perhaps as the saying goes, "the greater the ocean of knowledge, the wider the shore of our ignorance".

How true this cliché will hold as science and technology continues to help us reveal the mysteries behind our reality even as it helps us transcend beyond the confines of our humanity remains to be seen as the 2020s and the decades after them come.

ACKNOWLEDGEMENT

All thanks to the Creator of all things, who resides in light unapproachable. Whose only will for the universe is that all conscious intelligences would emerge to a life of immortality and share in the beauty of His glory.

To my Dad, who provided me a treasure trove of knowledge by allowing me unrestricted access to his library of books complete with a set of encyclopedia. As a curious kid growing up in a world and time of no computers or internet, this was my escape to worlds unknown.

To my wonderful family, the Idehen's, whose persistent love and patience I have often overstretched. Please remain patient with me; your love is the only assured warmth I have as I brave the many cold waves life often over floods us with.

To my family the Okebie's, whose support and affections provided me a safe haven for the writing of this book. To Patrick Okebie, a mentor, whose influence in my life remains indelible through the sands of time. To Grace Okebie, to whom I will forever be grateful to for care and for the gifts of Ray Kurzweil, Kai Fu Lee and Dan Brown's books.

I say a big thanks to my special friends Oluwaseun R. Olusegun, Precious Goggins and Uchenna Maduka who helped me with their sincere reviews of initial drafts.

To Scofield (Enyonmi), Didi (Big Dee), Cousin Shims and Ubi, Oneal Lajuwomi, Nnnena, Slyzee and Chuks, and to everyone who in one way or the other has helped to bend the will of fate to make this book possible, I am forever grateful. Thank you.

NOTES

INTRO

General Purpose Technology (GPTs): technologies that can affect an entire economy
https://en.wikipedia.org/wiki/General_purpose_technology

Elon Musk: 'With artificial intelligence we are summoning the demon.'
https://www.washingtonpost.com/news/innovations/wp/2014/10/24/elon-musk-with-artificial-intelligence-we-are-summoning-the-demon/

CHAPTER 1

Richard Buckminster Fuller: Life and Biography
https://en.wikipedia.org/wiki/Buckminster_Fuller

Knowledge Doubling Curve: Knowledge Doubling Every 12 Months, Soon to be Every 12 Hours
https://www.industrytap.com/knowledge-doubling-every-12-months-soon-to-be-every-12-hours/3950

Gordon Earle Moore: Life and Work
https://en.wikipedia.org/wiki/Gordon_Moore

Moore's Law:
http://www.mooreslaw.org/
https://www.britannica.com/technology/Moores-law
https://www.investopedia.com/terms/m/mooreslaw.asp

Raymond Kurzweil: Futurist and Inventor
https://en.wikipedia.org/wiki/Ray_Kurzweil

Technological singularity
https://en.wikipedia.org/wiki/Technological_singularity

The Singularity: Singularity: Explain It to Me Like I'm 5-Years-Old
https://futurism.com/singularity-explain-it-to-me-like-im-5-years-old

Can Futurists Predict the Year of the Singularity?
https://singularityhub.com/2017/03/31/can-futurists-predict-the-year-of-the-singularity/

Need For Quantum Computers, Satya Nadella: World Economic Forum
https://www.bbc.com/news/business-42797846

How blockchains could change the world, Don Tapscott:
https://www.mckinsey.com/industries/technology-media-and-telecommunications/our-insights/how-blockchains-could-change-the-world

Understanding Nanotechnology
https://www.explainthatstuff.com/nanotechnologyforkids.html

Grey goo and Nanotech
https://en.wikipedia.org/wiki/Gray_goo

New NVIDIA Research Creates Interactive Worlds with AI
https://nvidianews.nvidia.com/news/new-nvidia-research-creates-interactive-worlds-with-ai

The good, the bad and the ugly of 3D printing technology
https://www.ft.com/content/782461c8-9fe1-11e8-b196-da9d6c239ca8

The Ultimate Guide to 3D Printing
https://all3dp.com/3d-printing-3d-printer-guide-101-questions/

CHAPTER 2

The Law of Accelerating Returns, Ray Kurzweil
https://www.kurzweilai.net/the-law-of-accelerating-returns

2 Billion Jobs to Disappear by 2030, Thomas Frey
https://futuristspeaker.com/business-trends/2-billion-jobs-to-disappear-by-2030/

Alvin Toffler: American writer and futurist. Bio and life work
https://en.wikipedia.org/wiki/Alvin_Toffler

CHAPTER 3

Elon Musk's AI warning: Artificial Intelligence is a 'potential danger to the public'
https://www.express.co.uk/news/science/1204119/Elon-Musk-AI-warning-Artificial-Intelligence-danger-Neuralink-Elon-Musk-latest

Facebook–Cambridge Analytica data scandal
https://en.wikipedia.org/wiki/Facebook%E2%80%93Cambridge_Analytica_data_scandal

Google Photos labeled black people 'gorillas'
https://www.usatoday.com/story/tech/2015/07/01/google-apologizes-after-photos-identify-black-people-as-gorillas/29567465/

Sergey Brin: 2017 Founders' Letter, Alphabet
https://abc.xyz/investor/founders-letters/2017/

CHAPTER 4

100 Years of Progress
https://www.diamandis.com/blog/100-years-of-progress

Facebook Security Breach Exposes Accounts of 50 Million Users
https://www.nytimes.com/2018/09/28/technology/facebook-hack-data-breach.html

Twitter CEO and co-founder Jack Dorsey has account hacked
https://www.bbc.com/news/technology-49532244

2018 Google data breach
https://en.wikipedia.org/wiki/2018_Google_data_breach

Facebook's Zuckerberg takes broad lashing on Libra, 2020 election and civil rights at congressional hearing
https://www.washingtonpost.com/technology/2019/10/23/facebook-mark-zuckerberg-testifies-congress-election-libra/

Sundar Pichai had to explain to Congress why Googling 'idiot' turns up pictures of Trump
https://www.theverge.com/2018/12/11/18136114/trump-idiot-image-search-result-sundar-pichai-google-congress-testimony

The Enterprise AI Promise: Path to Value, SAS (PDF)
https://www.sas.com/content/dam/SAS/el_gr/doc/research1/ai-survey-2017.pdf

FBI Investigating Hate Site Linked to Accused Charleston Shooter
https://time.com/3929352/dylann-roof-website-manifesto/

Dylann Storm Roof, Charleston church shooting
https://en.wikipedia.org/wiki/Dylann_Roof

CHAPTER 5

Russia's first sex robot brothel opens ahead of World Cup in a bid to cash in on fans... and players
https://www.dailymail.co.uk/news/article-5713369/Russias-sex-robot-brothel-opens-ahead-World-Cup-bid-cash-fans-players.html

Burger-flipping robot begins first shift
https://www.bbc.com/news/av/technology-43292047/burger-flipping-robot-begins-first-shift

Burger-flipping robot begins first shift
https://www.cnbc.com/2019/11/22/self-driving-trucks-likely-to-hit-the-roads-before-passenger-cars.html

Jobs And Robots: 25 Countries Ranked On Job Loss Potential From Automation, Robotics, And AI
https://www.forbes.com/sites/johnkoetsier/2018/04/23/usa-ranks-9th-in-global-robotics-automation-job-loss-report-after-korea-germany-japan-canada/#1ec7ecd93f62

The Future of Jobs and Skills
https://reports.weforum.org/future-of-jobs-2016/chapter-1-the-future-of-jobs-and-skills/

Future of Jobs, Employment, Skills & Workforce Strategy for the 4th Industrial Revolution (PDF)
http://www3.weforum.org/docs/WEF_Future_of_Jobs.pdf

Robots will eliminate 6% of all US jobs by 2021,
https://www.theguardian.com/technology/2016/sep/13/artificial-intelligence-robots-threat-jobs-forrester-report

President J.F Kennedy, Nation's Manpower Revolution: Hearings Before the United States (Book)
Page 321,

CHAPTER 6

World's first AI news anchor unveiled in China
https://www.theguardian.com/world/2018/nov/09/worlds-first-ai-news-anchor-unveiled-in-china

10 Jobs That Are Safe in an AI World, Kai Fu Lee
https://medium.com/@kaifulee/10-jobs-that-are-safe-in-an-ai-world-ec4c45523f4f

AI Superpowers: China, Silicon Valley, and the New World Order (Book) – Kaai Fu Lee

A FUTURE THAT WORKS: AUTOMATION, EMPLOYMENT, AND PRODUCTIVITY (PDF)
https://www.mckinsey.com/~/media/mckinsey/featured%20insights/Digital%20Disruption/Harnessing%20automation%20for%20a%20future%20that%20works/MGI-A-future-that-works-Executive-summary.ashx

Why AI could destroy more jobs than it creates, and how to save them
https://www.techrepublic.com/article/ai-is-destroying-more-jobs-than-it-creates-what-it-means-and-how-we-can-stop-it/

CHAPTER 7

Responses to the Industrial Revolution
https://webs.bcp.org/sites/vcleary/ModernWorldHistoryTextbook/IndustrialRevolution/responsestoIR.html

The Original Luddites Raged Against the Machine of the Industrial Revolution
https://www.history.com/news/industrial-revolution-luddites-workers

Luddite, History
https://en.wikipedia.org/wiki/Luddite

Ada Lovelace Biography Computer Programmer, Mathematician
https://www.biography.com/scholar/ada-lovelace

The Future of Jobs and Skills: WEF report, Chapter 1
http://reports.weforum.org/future-of-jobs-2016/chapter-1-the-future-of-jobs-and-skills/

REALISING THE ECONOMIC AND SOCIETAL POTENTIAL OF RESPONSIBLE AI IN EUROPE (PDF)
https://www.accenture.com/_acnmedia/pdf-72/accenture-realising-economic-societal-potential-responsible-ai-2.pdf

Adapt or die: How to cope when the bots take your job
https://www.bbc.com/news/business-43259906

CHAPTER 8

Deloitte: Automation is here to stay...but what about your workforce? Preparing your organization for the new worker ecosystem
https://www2.deloitte.com/content/dam/Deloitte/global/Documents/Financial-Services/gx-fsi-automation-here-to-stay.pdf

McKinsey Global Institute: Retraining and reskilling workers in the age of automation
https://www.mckinsey.com/featured-insights/future-of-work/retraining-and-reskilling-workers-in-the-age-of-automation

New Generation Purpose (Millenial):The new rules of business leadership
http://business.edf.org/blog/tag/project-gigaton

Larry Fink's 2019 Letter to CEOs: Purpose & Profit
https://www.blackrock.com/corporate/investor-relations/larry-fink-ceo-letter

CHAPTER 9

Kai-Fu Lee's perspectives on two global leaders in artificial intelligence: China and the United States
https://www.mckinsey.com/featured-insights/artificial-intelligence/kai-fu-lees-perspectives-on-two-global-leaders-in-artificial-intelligence-china-and-the-united-states

Next Generation Artificial Intelligence Development PlanIssued by State Council
http://fi.china-embassy.org/eng/kxjs/P020171025789108009001.pdf

AI development plan draws map for innovation: China
https://www.chinadaily.com.cn/a/201908/05/WS5d476b48a310cf3e35563d0d.html

An Overview of National AI Strategies
https://medium.com/politics-ai/an-overview-of-national-ai-strategies-2a70ec6edfd

Tax robots and Universal Basic Income
https://techcrunch.com/2018/07/17/tax-robots-and-universal-basic-income/

Andrew Yang (Yang2020) What is the freedom dividend?
https://www.yang2020.com/what-is-freedom-dividend-faq/

Presidential candidate Andrew Yang will give $1,000 a month to 10 more families
https://www.cnbc.com/2019/09/12/andrew-yang-says-he-will-give-1000-a-month-ubi-to-10-more-families.html

History of basic income
https://basicincome.org/basic-income/history/

Why Did Hayek Support a Basic Income?
https://www.libertarianism.org/columns/why-did-hayek-support-basic-income

Martin Luther King's Economic Dream: A Guaranteed Income for All Americans https://www.theatlantic.com/business/archive/2013/08/martin-luther-kings-economic-dream-a-guaranteed-income-for-all-americans/279147/

Why Silicon Valley is embracing universal basic income
https://www.theguardian.com/technology/2016/jun/22/silicon-valley-universal-basic-income-y-combinator

CHAPTER 10

On the Origin of Species (Charles Darwin)
https://en.wikipedia.org/wiki/On_the_Origin_of_Species

ARE YOU LIVING IN A COMPUTER SIMULATION? Nick Bostrom
https://www.simulation-argument.com/simulation.html

Neuralink, a company in which Mr. Musk has invested $100 million
https://www.nytimes.com/2019/07/16/technology/neuralink-elon-musk.html

List of religious populations
https://en.wikipedia.org/wiki/List_of_religious_populations

What Is the Big Bang Theory?
https://www.space.com/25126-big-bang-theory.html

The Big Bang Theory: Singularity
https://en.wikiversity.org/wiki/The_Big_Bang_Theory

The Great Contrversy – Ellen G White Book.
https://m.egwwritings.org/en/book/12320.529

THE GALILEO AFFAIR
https://www.libraryofsocialscience.com/newsletter/posts/2016/2016-05-12-Galileo.html

Sam Harris TED Talk: Can We Build AI Without Losing Control Over It?
https://futureoflife.org/2016/10/07/sam-harris-ted-talk/?cn-reloaded=1

Sam Harris: Developing AI is Humanity "Building Some Sort of God"
https://futurism.com/sam-harris-developing-ai-is-humanity-building-some-sort-of-god

The AI Revolution: The Road to Superintelligence
https://waitbutwhy.com/2015/01/artificial-intelligence-revolution-1.html

The AI Revolution: Our Immortality or Extinction
https://waitbutwhy.com/2015/01/artificial-intelligence-revolution-2.html

How Far Are We From Achieving Artificial General Intelligence?
https://www.forbes.com/sites/cognitiveworld/2019/06/10/how-far-are-we-from-achieving-artificial-general-intelligence/#63c7cfd26dc4

China's Tianhe-2 is still the world's fastest supercomputer, but Cray is on a resurgence
https://www.extremetech.com/extreme/218078-chinas-tianhe-2-is-still-the-worlds-fastest-supercomputer-but-cray-is-on-a-resurgence

Thinking Inside the Box:Controlling and Using an Oracle AI
https://www.nickbostrom.com/papers/oracle.pdf

CHAPTER 11

Way of the Future: A New Church Worships an AI God
One day, super intelligences may be advanced enough to be considered divine.
https://futurism.com/way-future-new-church-worships-ai-god
Way of the Future Church: Humans United in support of AI, committed to peaceful transition to the precipice of consciousness.

http://www.wayofthefuture.church/

Anthony Levandowski, Church worship of an Artificial Intelligence
https://www.businessinsider.com/anthony-levandowski-way-of-the-future-church-where-people-worship-ai-god-2017-11?IR=T

The Allegory of the Cave From the Republic of Plato, Plato's Best-Known Metaphor About Enlightenment
https://www.thoughtco.com/the-allegory-of-the-cave-120330

Cogito, ergo sum: René Descartes, Discourse on the Method
https://en.wikipedia.org/wiki/Cogito,_ergo_sum

3 Superb arguments for why we live in a matrix – and 3 arguments that refute them
https://bigthink.com/arguments-live-matrix--arguments-against?rebelltitem=4#rebelltitem4

We might live in a computer program, but it may not matter
http://www.bbc.com/earth/story/20160901-we-might-live-in-a-computer-program-but-it-may-not-matter

Bank of America thinks there's a 50% chance we live in a Matrix
https://www.businessinsider.com/bank-of-america-thinks-theres-a-50-chance-we-live-in-a-matrix-2016-9?IR=T

Is the Universe a Simulation? Scientists Debate
https://www.space.com/32543-universe-a-simulation-asimov-debate.html

Are We Living in a Computer Simulation?
https://www.pbs.org/wgbh/nova/article/are-we-living-in-a-computer-simulation/

Quantum Theory: Part of the Einstein exhibition.
https://www.amnh.org/exhibitions/einstein/legacy/quantum-theory

Quantum Behavior: The Richard Feynman Lectures
https://www.feynmanlectures.caltech.edu/III_01.html

USEFUL READINGS ON CHEMISTRY: Quantum Mechanics, by Richard P. Feynman
https://bouman.chem.georgetown.edu/general/feynman.html

Quantum mechanics: From Wikipedia, the free encyclopedia
https://en.wikipedia.org/wiki/Quantum_mechanics

It from bit? How come existence?
https://plus.maths.org/content/it-bit

John Wheeler's Participatory Universe
https://futurism.com/john-wheelers-participatory-universe

The Holy Bible, KJV: Let there be light
https://biblehub.com/genesis/1-3.htm

Nobel-Winning Physicist Niels Bohr on Subjective vs. Objective Reality and the Uses of Religion in a Secular World
https://www.brainpickings.org/2018/02/01/niels-bohr-science-religion/

CHAPTER 12

The Creation of Adam: Michelangelo
https://en.wikipedia.org/wiki/The_Creation_of_Adam

God Created: Seventh Day Adventist Sabbath School
https://www.sabbathschoolpersonalministries.org/assets/sspm/Lessons/2019/Q3/English/Teacher/ABSG-19-Q3-TE-L01.pdf

Chapter 2: The Creation, The Story of Patriarchs and Prophets,
https://www.ellenwhite.info/books/ellen-g-white-book-patriarchs-and-prophets-pp-2.htm

Transhumanism And The Future Of Humanity: 7 Ways The World Will Change By 2030
https://www.forbes.com/sites/sarwantsingh/2017/11/20/transhumanism-and-the-future-of-humanity-seven-ways-the-world-will-change-by-2030/#6c058e917d79

Transhumanism
https://en.wikipedia.org/wiki/Transhumanism

Robert Downey Jr to Return as Iron Man in Disney+
https://www.thewrap.com/robert-downey-jr-iron-man-in-what-if/

Advancing Nature and Technology for a better Future
https://humanityplus.org/

Blue Brain project finds how neurons form billions of synaptic connections
https://www.healtheuropa.eu/blue-brain-synaptic-connections/93112/

Nikolai Fyodorovich Fyodorov
https://en.wikipedia.org/wiki/Nikolai_Fyodorovich_Fyodorov

Mormon Transhumanist Organisation: As disciples of Jesus Christ, we believe in using technology to serve, lift, and love.
https://transfigurism.org/

Christian Transhumanist Association: Could Science & Faith work together to create a radically better Future?
https://www.christiantranshumanism.org/

Transhumanist Religions 2.0 with Giulio Prisco

https://www.kurzweilai.net/transhumanist-religions-2-0-with-giulio-prisco

Introduction to Transhumanism, Frank Tipler: Simulation and Cosmism

https://www.academia.edu/9505954/Introduction_to_Transhumanism

The Omega Point and Beyond: The Singularity Event

http://www.ajnr.org/content/33/3/393

Pierre Teilhard de Chardin: The Omega Point

https://en.wikipedia.org/wiki/Omega_Point

ABOUT THE AUTHOR

Author Full Name: Kelly, Osahen Idehen.

Kelly Idehen is an emerging technology thought leader, entrepreneur, AI researcher and influencer for digital innovation.

He is passionate about the applications and impacts of disruptive technologies for solving real life problems faced by people especially in underdeveloped countries.

He is currently founder of LearnHub.Africa and the CEO of LinkOrion Technology where he consults for brands and helps them undergo the process of Digital Transformation.

When not fully immersed in research, speaking engagement and technology related projects, Kelly loves to spend his time on social impact projects, helping young people through Digital skill programs, STEM education and mentorship; as a commitment to his mission of getting more people ready for a future that will be completely disrupted by technology.

For more interaction with Kelly Idehen,
Visit his website – www.kellyidehen.com
Connect with him on Linkedin – Kelly Idehen (@Iconickelx)
Twitter and Instagram – @ Iconickelx